Smart Sensors, Measurement and Instrumentation 5

Series Editor

Subhas Chandra Mukhopadhyay
School of Engineering and Advanced Technology (SEAT)
Massey University (Turitea)
Palmerston North
New Zealand
E-mail: S.C.Mukhopadhyay@massey.ac.nz

For further volumes:
http://www.springer.com/series/10617

Subhas Chandra Mukhopadhyay

Intelligent Sensing, Instrumentation and Measurements

Springer

Prof. Subhas Chandra Mukhopadhyay
School of Engineering and Advanced Technology
Massey University (Turitea Campus)
Palmerston North
New Zealand

ISSN 2194-8402 ISSN 2194-8410 (electronic)
ISBN 978-3-642-37026-7 ISBN 978-3-642-37027-4 (eBook)
DOI 10.1007/978-3-642-37027-4
Springer Heidelberg New York Dordrecht London

Library of Congress Control Number: 2013933100

Printed on acid-free paper

Springer is part of Springer Science+Business Media (www.springer.com)

Dedicated
to
Mrs. Chandra Jayasundera

Preface

In recent times, the wireless sensors and sensors network have been widely used in many applications such as monitoring environmental parameters, monitoring and control of industrial situations, intelligent transportation, structural health monitoring, health care and so on. The advancement of electronics, embedded controller, smart wireless sensors, networking and communication have made it a possibility of the development of a low cost, low power smart wireless sensors nodes. Though there are many publications on the topic, the book will provide very useful information and basic knowledge to develop a wireless sensor network from scratch.

The complete book is divided into seven chapters.

The fundamental of sensors have been described in Chapter 1. A short descriptions of a few sensors used for making wireless sensors or smart sensors has been provided. For implementing a smart sensor the basic knowledge of this chapter is very important.

For making a smart sensor or a wireless sensor network, the sensor is connected to a processor. There is always some issues of interfacing a sensor to a processor. The chapter 2 has described some issues of interfacing sensors to a processor and signal conditioning.

The Chapter 3 has introduced some fundamentals of developing wireless sensor network (WSN). The importance of developing of WSN and different components to achieve it has been explained.

The basics of power supply and different energy harvesting techniques for the sensor node in WSN has been explained in chapter 4.

The chapter 5 provides the basic knowledge to implement a WSN. The emphasis has been given on ZigBee based system design. The readers will get complete idea of implementing the configuration of a sensor node and the coordinator. The chapter then explains the programming details of making a GUI for data reception and storage.

A few techniques of analysing the sensors data has been explained in the chapter 6.

The chapter 7 has described the implementation of three projects involving wireless sensors and sensors network. The readers will get a very good idea about the details of implementation.

I am very much indebted to many of my students (past and present) who actually contributed significantly on the manuscript. It would be unfair if I do not mention their names (not in order): Nagender Kumar Suryadevara, Quan Vu, Sean Kelly, Gerard Mendez, Mathias Haefke, Chinthaka Gooneratne, Michael

Sutherland, Anuroop Gaddam, Mohd Syaifudin Abdul Rahman, Vishnu Kasturi, Karan Singh Malhi, Matias Haefke, Julia Schnepper, Satinder Singh Gill, Mohd. Amri Yunus, Chagitha Ranhotigamage, Hatim Al Abri. I would like to specially mention Nagender Kumar Suryadevara who has contributed a huge part of the chapters 5 and 6. I am also grateful to many researchers for their technical papers, many of which have not been mentioned. I also like to express my thanks to my wife, Krishanthi and my children, Sakura and Hiroshi, for their continuous support and help.

Subhas Chandra Mukhopadhyay
School of Engineering and Advanced Technology
Massey University (Manawatu)
Palmerston North, New Zealand

Contents

Chapter 1
Sensors Fundamental

Introduction

In our everyday life we use different sensors for various applications. A sensor is a device or a part of a system which is used for the measurement of physical, chemical, biological or any other parameters. Usually, a sensor gives an equivalent signal either in voltage or in current form which can be measured, processed, stored and transmitted. For example, a Resistance Temperature Detector (RTD) shows change of resistance with the change of temperature. The change of resistance is measured for the measurement of temperature. Using the RTD in a suitable electronic circuit, an equivalent voltage signal can be obtained as a function of temperature.

Ideally, it is expected that the sensor should only be sensitive to the parameter it measures and should not be influenced by any other parameters in the surroundings. Also, the measured parameters should not influence the sensors.

The sensor changes its output with the change of the measured parameters. The relationship between the sensor's output and the measured parameter is defined as the sensitivity of the sensors. For example, in a tacho-generator the output is a voltage output in volt (V) which is a function of speed in RPM (Revolutions per Minute). So the sensitivity is expressed as V/RPM. In an ideal situation the sensors should provide linear relationship between the output and the measured parameter. But, in many situations, this may not be true and the output can be mathematically expressed as a function of the measured parameter.

The sensor technologies have made an enormous impact on the modern day industries. There are many sensors available in the market and a lot of researches are currently undergoing to develop new, smart, efficient and high performance sensors. In the following section the classification of sensors are discussed.

1.1 Sensor Classification

Figure 1.1 shows a few sensors developed for some new applications at Massey University, New Zealand. Figure 1.2 shows some sensors which are readily available in the market. A few pictures are downloaded from internet without any

S.C. Mukhopadhyay: *Intelligent Sensing, Instrumentation and Measurements*, SSMI 5, pp. 1–27.
DOI: 10.1007/978-3-642-37027-4_1

reference details. There are varieties of sensors available in the markets that are ready to be used in a sensing system. In this particular section we will be looking at some of the sensor technologies that are available on the market that could be used for making a sensing system. The operating principles as well as their advantages and disadvantages will be discussed.

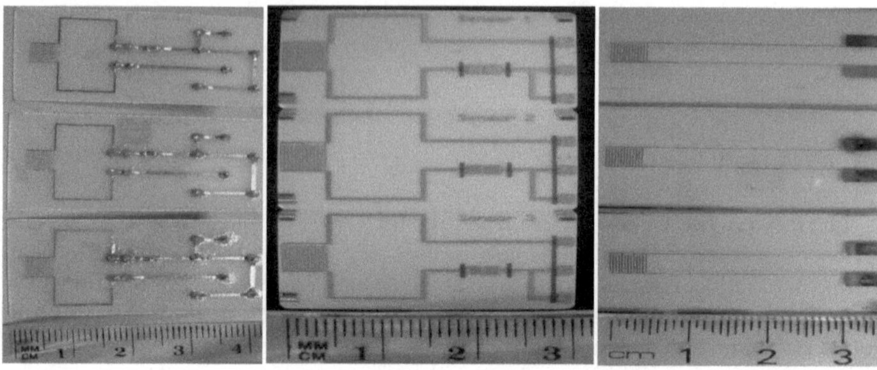

Fig. 1.1 Research on new types of sensors under development at Massey University, NZ

Fig. 1.2 Sensors available off the shelf

Table 1.1 shows some of the sensors along with their measurand (defined as a physical quantity, property or a condition) and transduction principle. Various methods are utilised to measure different phenomenon and consequently they are used in respective applications.

Table 1.1 Measurand and transduction principle of a few sensors

Measurand	Transduction principle
Temperature	Thermistor, Thermocouple, Thermo-mechanical
Humidity	Resistive, capacitive
Pressure	Piezo-resistive
Flow	Pressure change, thermistor
Force	Piezoelectric, piezoresistive
Torque	Piezoresistive, optoelectronic
Strain	Piezoresistive
Vibration	Piezoresistive, piezoelectric, optical fiber, sound, ultrasound
Position	E-mag, GPS, contact sensor
Velocity	Doppler, Hall effect, optoelectronic
Angular velocity	Optical encoder
Acceleration	Piezoresistive, piezoelectric, optical fiber
Proximity	Hall effect, capacitive, magnetic, seismic, acoustic, RF
Tactile/contact	Contact switch, capacitive

1.2 Thermal Sensors

Thermal sensors are used mainly for temperature measurement. Thermal sensor is one of the most commonly used sensors in the modern world. There are five different families of thermal sensors available, each family of thermal sensors has its own advantages and disadvantages. This section provides an overview of these thermal sensors.

1.2.1 Thermistors

Thermistors are widely used for measurements of temperature in a simple manner. A thermistor is a temperature-sensing element that is composed of sintered semiconductor material which exhibits a large change in resistance proportional to a small change in temperature. Thermistors usually have negative temperature coefficients which mean that the resistance of the thermistor decreases with the increase in temperature. Thermistors are not very accurate or stable but they are easier to wire, cost less and are very popular in automation panel as they are used directly. Figure 1.3 shows some commonly available thermistors in the market.

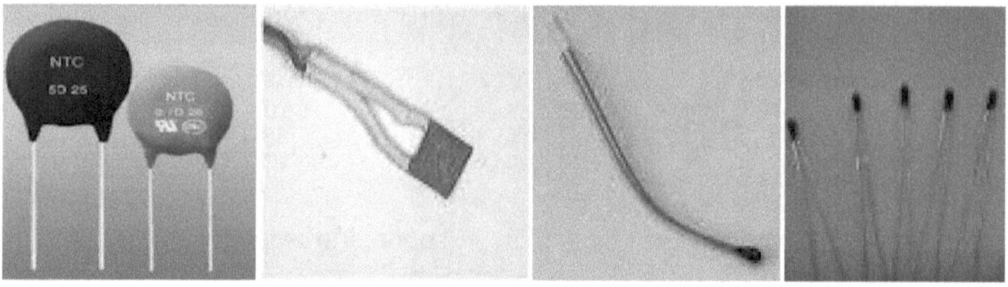

Fig. 1.3 A few commonly available thermistors

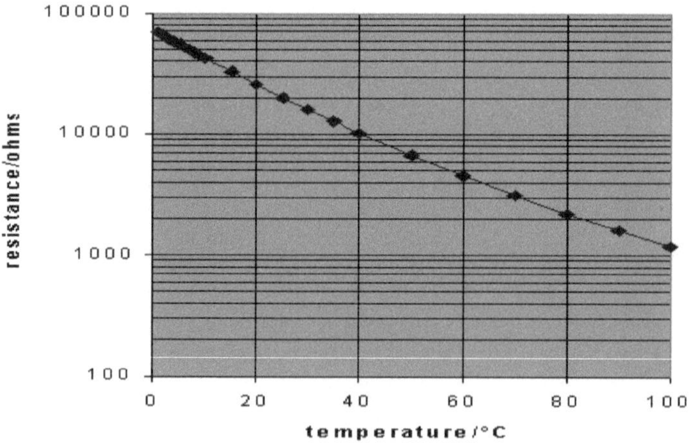

Fig. 1.4 A typical resistance-temperature characteristic of a thermistor

Figure 1.4 shows a typical resistance-temperature characteristic of a thermistor. The advantages of thermistors are: (i) highly sensitive and stable, (ii) cheap and (iii) quite accurate over small temperature ranges. But they have a few disadvantages too. The disadvantages are: (i) Non-linear resistance-temperature characteristics, (ii) limited temperature operating range and (iii) self heating.

1.2.2 Thermocouple

A thermocouple consists of two dissimilar metals, joined together at one end. When the junction of the two metals is heated or cooled a voltage is produced that can be correlated back to the temperature. The representation of the principle of operation of a thermocouple sensor is shown in figure 1.5. Thermocouples are suitable for measuring over a large temperature range. They are not very much suitable for applications where smaller temperature differences need to be measured with high accuracy. For such applications thermistors and resistance temperature detectors are more suitable. Applications include temperature measurement for kilns, gas turbine exhaust, diesel engines, and other industrial processes. Figure 1.6 shows a few thermocouple sensors available off-the-shelf. Table 1.2 shows the range and sensitivity of the commonly available thermocouple sensors. Figure 1.7 shows the sensitivity of the different types of thermocouple sensors.

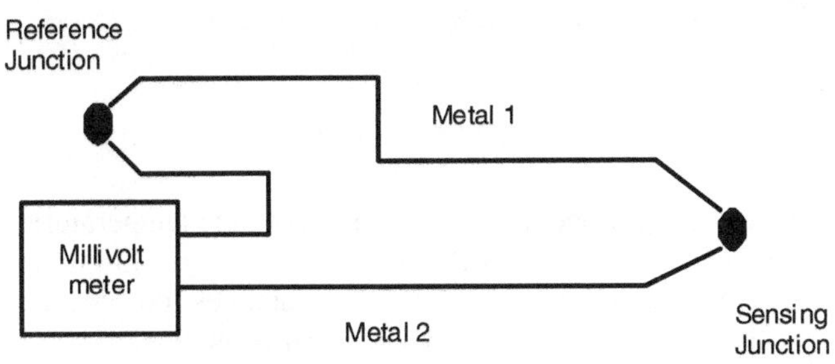

Fig. 1.5 The principle of operation of a thermocouple sensor

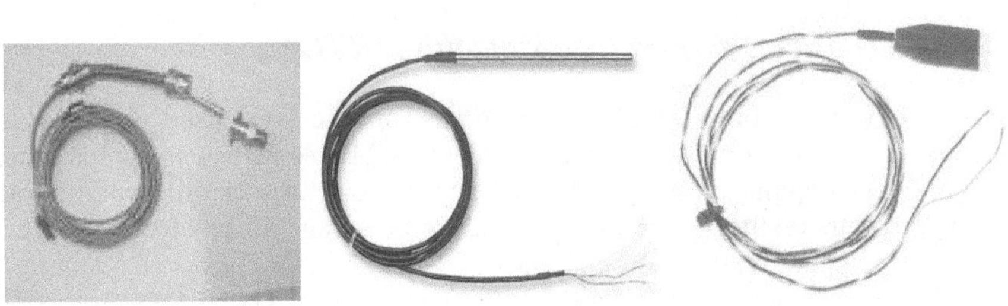

Fig. 1.6 A few thermocouple sensors available off the shelf

Table 1.2 Material characteristics of Thermocouples

Ref.	Materials	Range (°C)	Sensitivity (μV/°C)
E	Chromel / Constantan	-200 to 1000	63
J	Iron / Constantan	-200 to 900	53
T	Copper / Constantan	-200 to 400	43
K	Chromel / Alumel	-200 to 1300	41
R	Platinum / Platinum 13% rhodium	0 to 1400	6

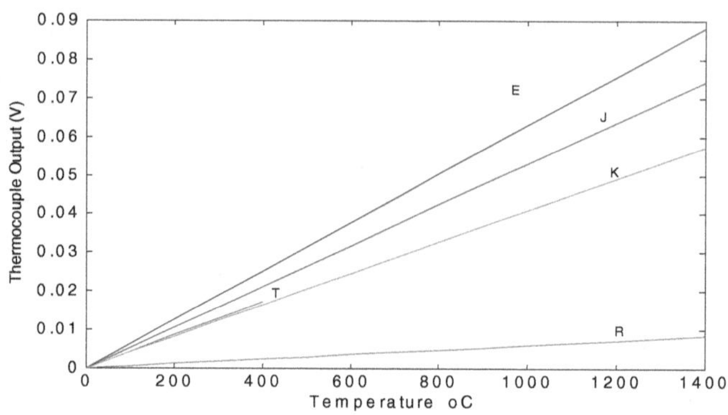

Fig. 1.7 Characteristics of some thermocouple sensors

A few advantages of thermocouple sensors are: (i) wide temperature range, 230 °C – 2300 °C, (ii) relatively cheap, (iii) highly accurate, (iv) low long-term drift and (v) fast response time. There are a few disadvantages too, which are: (i) the relationship between the temperature and the thermocouple signal is non-linear, (ii) they provide a low output signal (μV), (iii) they are vulnerable to corrosion and (iv) calibration of thermocouples can be tedious and difficult.

1.2.3 Resistance Temperature Detectors (RTD)

Resistance Temperature Detectors or RTDs are sensors used to measure temperature by correlating the resistance of the RTD element with temperature. RTDs are very popular in many industrial applications such as: air conditioning, food processing, textile production, plastics processing, micro-electronics, and exhaust gas temperature measurement. Most RTD elements consist of a length of fine coiled wire wrapped around a ceramic or glass core. The element is usually quite fragile, so it is often placed inside a sheathed probe to protect it. Figure 1.8 shows a few RTDs available off-the-shelf.

A few advantages of RTD sensors are: (i) Linear over wide temperature operating range, (ii) Relatively accurate and (iii) Good stability and repeatability at high temperature (65 - 700 °C). The disadvantages are: (i) Low sensitivity, (ii) Higher cost compared to thermocouples and (iii) vulnerable to shock and vibration.

Fig. 1.8 Off-the-shelf Resistance Temperature Detectors

Figure 1.9 shows different connection diagrams by which a RTD sensor can be connected to measure temperature. The output of the Wheatstone bridge can be interfaced to a processor for the calculation of temperature. (Source: http://en.wikipedia.org/wiki/Resistance_thermometer). The connection diagrams shown are applicable to other temperature sensors too.

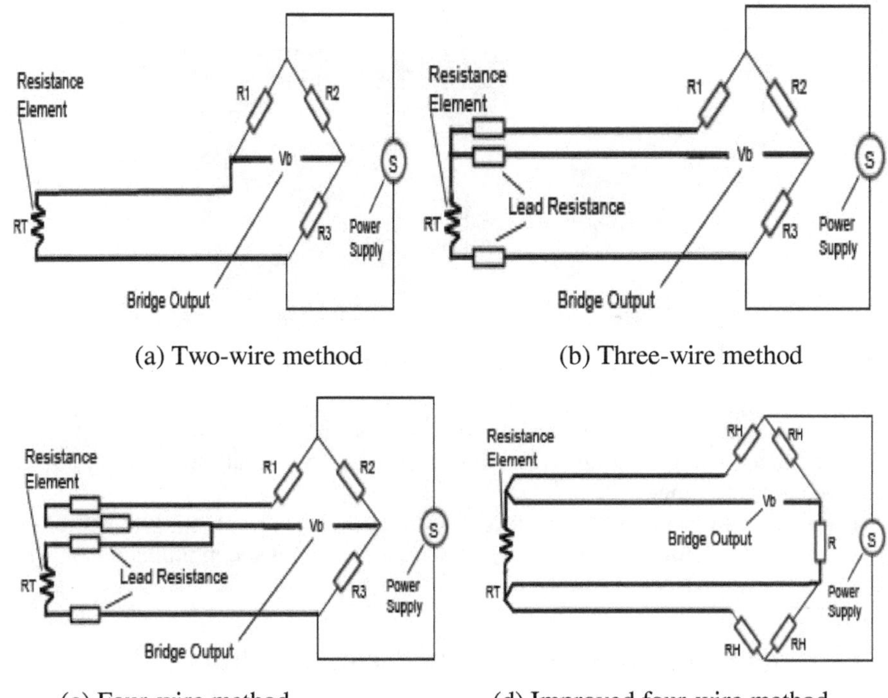

(a) Two-wire method (b) Three-wire method

(c) Four-wire method (d) Improved four-wire method

Fig. 1.9 Different methods of connection of RTDs

1.3 Humidity Sensors

A humidity sensor measures the humidity level in the air. The humidity is one the important weather parameter and needs to be measured regularly. It is also sometimes necessary to measure humidity in homes for people as the humidity gets affected as part of home heating, ventilating, and air conditioning systems. Humidity sensors can also be used in cars, office and industrial systems, and in meteorology stations to report and predict the weather. There are three types of humidity sensors: resistive, capacitive, and thermal conductivity humidity sensor. The following section discusses their operating principles as well as addresses their advantages and disadvantages.

1.3.1 Resistive Humidity Sensors (RHS)

Resistive Humidity Sensors (RHS) measure the changes in electrical impedance of a hygroscopic medium such as conductive polymer, salt, or treated substrate. These sensors are suitable for use in control and display products for industrial, commercial, and residential applications. RHS consists of noble metal electrodes either deposited on a substrate by photo-resist techniques or wire-wound electrodes on a plastic or glass cylinder. Figure 1.10 shows a few typical resistive humidity sensors.

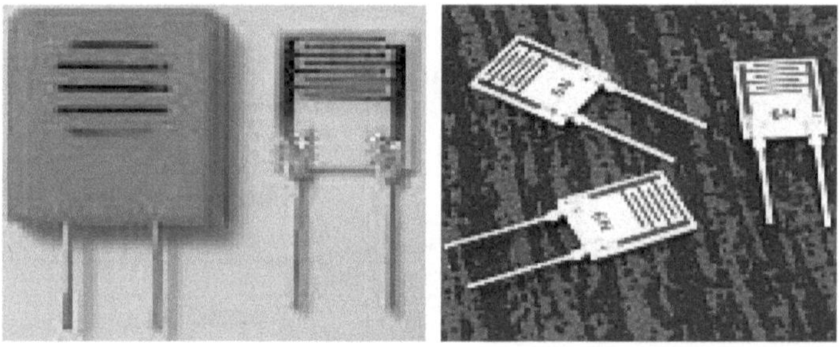

Fig. 1.10 Resistive humidity sensor

A few advantages of resistive humidity sensor are: (i) Fast response time, (ii) Almost linear voltage output, (iii) High accuracy, (iv) Small size and (v) Wide relative humidity range.

There are a few disadvantages too. They are: (i) Lower operating temperature, (ii) Sensitive to chemical vapors, (iii) Low tolerance against contaminants and (iv) Low condensation tolerance.

1.3.2 Capacitive Humidity Sensors (CHS)

Capacitive Humidity Sensors (CHS) are widely used in industrial, commercial, and weather telemetry applications. CHS consists of a substrate on which a thin film of polymer or metal oxide is deposited between two conductive electrodes. The sensing surface is coated with a porous metal electrode to protect it from contamination and exposure to condensation. The substrate is typically glass, ceramic, or silicon. The changes in the dielectric constant of a CHS are nearly directly proportional to the relative humidity of the surrounding environment. Figure 1.11 shows a few typical capacitive humidity sensors.

Fig. 1.11 Capacitive humidity sensors

The advantages of the Capacitive Humidity Sensors are: (i) ability to function in high temperatures environments (up to 200°C), (ii) Near linear voltage output, (iii) Wide RH range, (iv) High condensation tolerance, (v) Reasonable resistance to chemical vapors and contaminants, (vi) Minimal long-term drift, (vii) High accuracy and (viii) Small in size and low cost.

The disadvantages are: (i) Limited sensing distance and (ii) Sensor integration can be tedious and difficult.

1.3.3 Thermal Conductivity Humidity Sensors (TCHS)

Thermal Conductivity Humidity Sensors (TCHS) measure the absolute humidity by quantifying the difference between the thermal conductivity of dry air and that of air containing water vapor. These sensors are suitable for applications such as: paper, cooking, chemical solids, kilns for drying wood, machinery for drying textiles, pharmaceutical production and food dehydration. TCHS consists of two matched negative temperature coefficient (NTC) thermistor elements in a bridge circuit: one is hermetically encapsulated in dry nitrogen and the other is exposed to the environment. A few typical Thermal Conductivity Humidity Sensors are shown in figure 1.12.

The advantages of the Thermal Conductivity Humidity Sensors are: durable, has the ability to operate at high temperature up to 600°C, has excellent immunity to chemical and physical contaminants, provide high accuracy and high condensation tolerances. The disadvantages are: it may respond to any gas that has thermal properties different from those of dry nitrogen and is expensive.

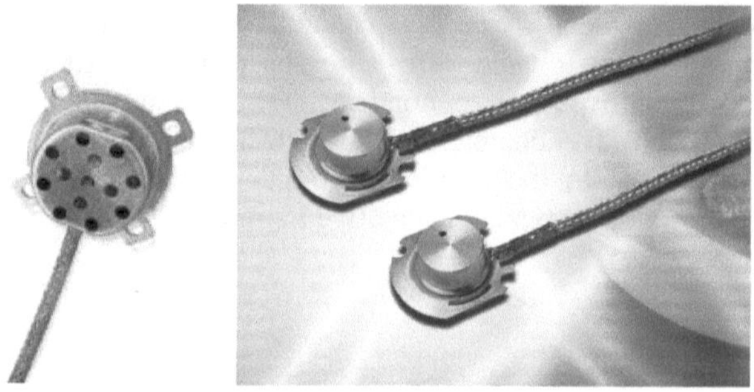

Fig. 1.12 Thermal conductivity humidity sensor

1.4 Capacitive Sensors

The capacitive sensors are very popular in many different applications, mainly to measure the dielectric properties. The parallel plate capacitor has two electrodes as shown in figure 1.13a and the electrodes are connected to a voltage source. The necessary electrical connection to measure the capacitance is shown in figure 1.13b. The capacitors are formed between positive and negative electrodes. The capacitance of a parallel plate capacitor is given by:

$$C = \frac{\varepsilon_0 \varepsilon_r A}{d} \tag{1.1}$$

Where, C is the capacitance in Farad

ε_0 is the absolute permittivity of free space, $\varepsilon_0 = 8.854 * 10^{-12}$ F/m

ε_r is the relative permittivity of the material.

A is the effective area.

d is the effective separation distance between the positive and the negative electrode.

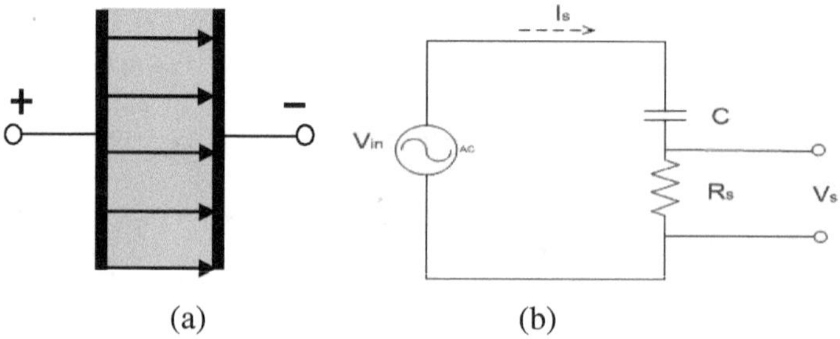

Fig. 1.13 (a) A simple parallel plate capacitor; (b) the electrical connection for parameter measurement

With respect to the figure 1.13b, the impedance of the capacitive sensor can be calculated by:

$$Z = \frac{V_{in}}{I_s} = \frac{V_{in}}{V_s / R_s} \quad \text{Or, } Z = \frac{V_{in}}{V_s} * R_s \tag{1.2}$$

Where, V_{in} is the input or excitation voltage across the system involving the capacitive sensor and a series resistance, R_s. The series resistance should be properly chosen so that it doesn't affect the measurement.

I_s is the supply current from the source.

V_s is the voltage across the series resistance, R_s.

Both the magnitude and the phase of the sensor impedance are measured. The real part (R) and the imaginary part (capacitive reactance, X_c) of the sensor is given by;

$$R = Z\cos\theta - R_s \tag{1.3}$$

$$X_c = Z\sin\theta \tag{1.4}$$

Depending on the materials, both the real and imaginary parts can be used to determine the properties of the system under investigation. From equation (1.4) the effective capacitance can be calculated by:

$$C = \frac{1}{2\pi f X_c} \tag{1.5}$$

1.5 Planar Interdigital Sensors

The operating principle behind the planar interdigital sensor is very similar to the one observed in a parallel plate capacitor. Figure 1.14 shows the relationship between a parallel plate capacitor and an interdigital sensor, and how the transition occurs from the capacitor to a sensor. There is an electric field between the positive and negative electrodes and figures 1.14a, b and c show how these fields pass through the material under test (MUT). Thus material dielectric properties as well as the electrode and material geometry affect the capacitance and the conductance between the two electrodes.

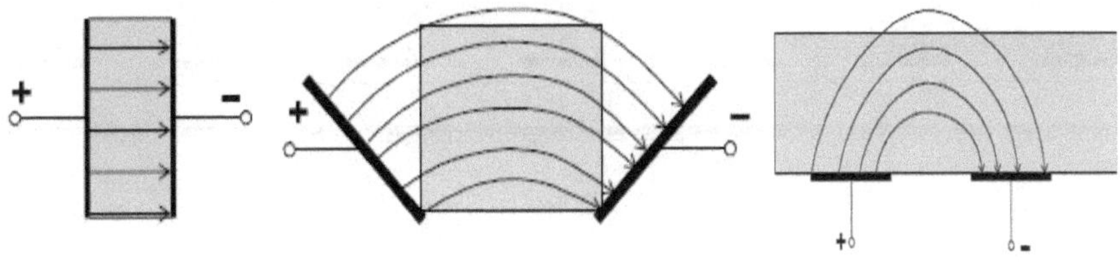

Fig. 1.14 (a) a parallel plate capacitor; (b) and (c) an interdigital sensor to provide a one sided access to the material under test

The electrodes of an interdigital sensor are coplanar. Hence, the measured capacitance will have a very low signal-to-noise ratio. In order to get a strong signal the electrode pattern can be repeated many times. This leads to a structure known as an interdigital structure. The term "interdigital" refers to a digit-like or finger-like periodic pattern of parallel in-plane electrodes, used to build up the capacitance associated with the electric fields that penetrate into a material sample. This is shown in figure 1.15. In figure 1.15 one set of electrodes are connected or driven by an AC voltage source while the other set are connected to ground. An electric field is formed between the driven and the ground electrodes. The figure 1.16 shows the electric field lines for different spacing of the electrodes. It can be seen that the depth of penetration of the electric field lines vary for different wavelengths. The wavelength (λ) of interdigital sensors is the distance between two adjacent electrodes of the same type. In figure 1.16 there are three lengths (l_1, l_2 and l_3) showing the different penetration depths with respect to the wavelengths of the sensor.

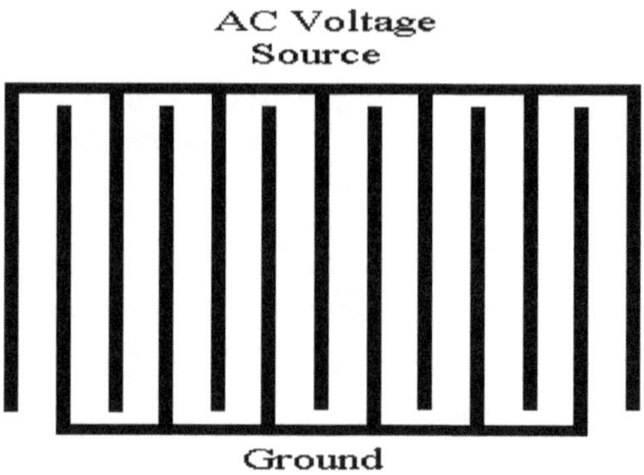

Fig. 1.15 Interdigital sensor structure, where the electrodes follow a finger-like or digit-like pattern

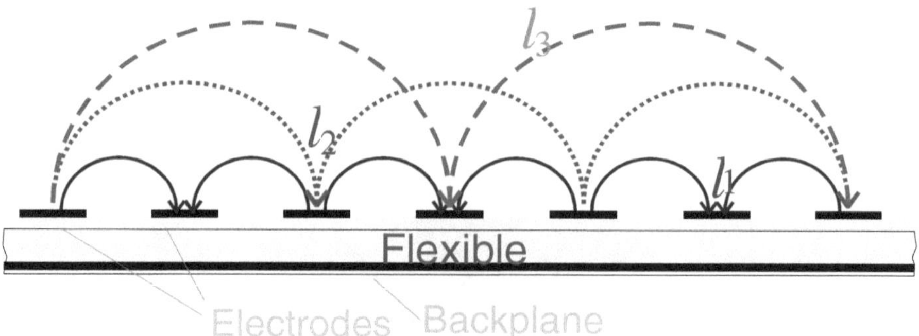

Fig. 1.16 Electric field formed between positive and negative electrodes for different pitch lengths, (l_1, l_2 and l_3) [33]

Planar interdigitated array electrodes have many applications such as gas detection, determining components in aqueous solutions, estimation of fiber, moisture and titanium dioxide in paper pulp and complex permittivity characterization of materials.

1.6 Planar Electromagnetic Sensors

A planar electromagnetic sensor is based on a combination of planar interdigital sensor and a planar meander sensor. The sensor creates a high-frequency electric and magnetic field driven by an alternating supply. By measuring the modification of the electromagnetic field which is caused by the surrounding material, the properties of the system under test are estimated. The planar electromagnetic sensor system can detect not only metallic and/or non-metallic materials but it can detect the material properties of a mixture of conducting, magnetic and dielectric materials. The picture of one of the sensors is shown in the figure 1.17. The sensor is connected to an external function generator which provides an alternating 10 Volt peak–to-peak (or any suitable magnitude) sine waveform signal between point 1 and point 4 as shown in the top view in figure 1.17. The bottom layer shows the sensor consists a spiral meander planar coil surrounding (each turn is distanced at 0.5 mm) an interdigital sensor which connected in series. The combined meander coil and the interdigital coil are connected to an alternating supply voltage. An alternating electromagnetic field is generated around the whole sensor due to the alternating electric current. Impedance values from meander and interdigital sensors are used as the characterization parameters of the material under test.

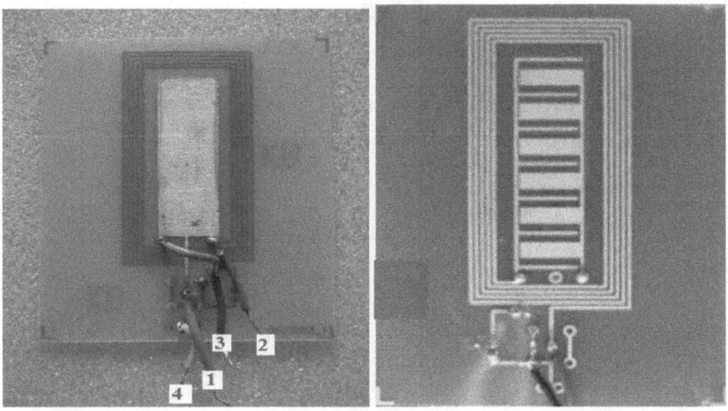

Fig. 1.17 Planar Electromagnetic sensor: Left-hand side: top view and right hand side: bottom view

The equivalent circuit can be seen from figure 1.18. The sensor is connected to a function generator where R_g is the output resistance with a nominal value of 50 Ω. R_{is} and X_{is} represent the real part and the imaginary part of the interdigital sensor, R_{ms} and X_{ms} represent the real part and the imaginary part of the meander sensor and R_1 represent series surface mount series resistor connected to the sensor.

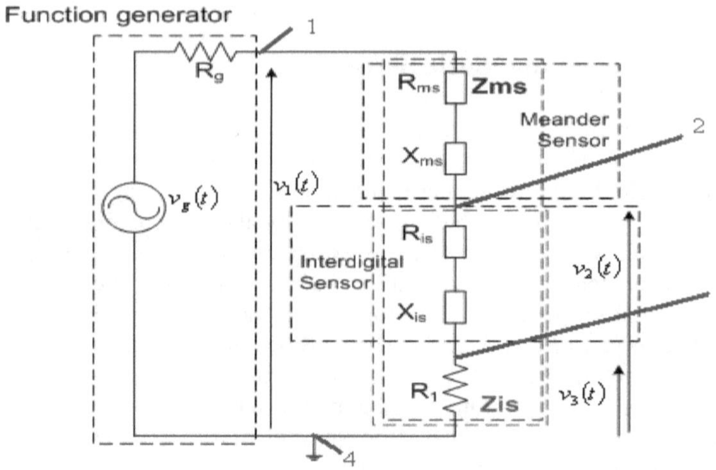

Fig. 1.18 An electrical equivalent circuit of the inductive-capacitive planar sensors connected in series

The calculation of the equivalent circuit parameters are done as follows:
The total impedance Zs is given by,

$$Z_s = \frac{V_1 \angle \theta}{V_3 \angle 0} R_1 \tag{1.6}$$

where θ is the phase difference between $V_1(t)$ with $V_3(t)$ in degree, taking $V_3(t)$ as reference. $V_3(t)$ is the rms voltage across the surface mount resistor (between point 3 and point 4). $V_3(t)$ is considered as the reference so that its phase angle is 0°.

For the interdigital sensor, we can have the following:

$$Z_{is} = \frac{V_2 \angle \theta_2}{V_3 \angle 0^\circ} R_1 \tag{1.7}$$

where V_2 is the rms values of voltage between point 2 and point 4. θ_2 is the phase difference between $v_2(t)$ with $v_3(t)$ in degree. Therefore, R_{is} and X_{is} are given by:

$$R_{is} = |Z_{is}| \cos(\theta_2) - R_1 \tag{1.8}$$

$$X_{is} = |Z_{is}| \sin(\theta_2) \tag{1.9}$$

R_{ms} and X_{ms} indicate the real part and the imaginary part of the meander sensor, respectively.

The impedance related to meander part R_{ms}, X_{ms}, and Z_{ms} are given by:

$$R_{ms} = |Z_1|\cos(\theta_1) - R_{is} - R_1 \tag{1.10}$$

$$X_1 = |Z_1|\sin(\theta_1) \tag{1.11}$$

$$X_{ms} = X_1 - X_{is} \tag{1.12}$$

$$Z_{ms} = R_{ms} + jX_{ms} = \sqrt{(R_{ms})^2 + (X_{ms})^2}\angle\tan^{-1}\left(\frac{X_{ms}}{R_{ms}}\right) = |Z_{ms}|\angle\theta_3 \tag{1.13}$$

The suggested operating frequency depends on the material properties and may be different for different materials. The first experiment involved establishing the reference values by making the measurement of all real part and imaginary part (interdigital and meander) of the sensors. For measurement related to water contamination, distilled water/mili-q water may be used as a reference.

Afterwards, the experiments are repeated with any water samples to measure the sensitivities given as the following:

$$\%\text{Real\_part\_interdigi} = \sum_f \left[\frac{(R_{is})_{sample} - (R_{is})_{distilled}}{(R_{is})_{distilled}} \times 100\right] \tag{1.14}$$

$$\%\,\text{Imaginary\_part\_interdigi} = \sum_f \left[\frac{|(X_{is})_{sample} - (X_{is})_{distilled}|}{|(X_{is})_{distilled}|} \times 100\right] \tag{1.15}$$

$$\%\text{Real\_part\_meander} = \sum_f \left[\frac{(R_{ms})_{sample} - (R_{ms})_{distilled}}{(R_{ms})_{distilled}} \times 100\right] \tag{1.16}$$

$$\%\text{Imagi\_part\_meander} = \sum_f \left[\frac{|(X_{ms})_{sample} - (X_{ms})_{distilled}|}{|(X_{ms})_{distilled}|} \times 100\right] \tag{1.17}$$

Where $(R_{xx})_{distilled}$ is the real part when the sensor is immersed in the distilled water/mili-q water and $(R_{xx})_{sample}$ is the real part of the impedance value when the sensor is immersed in the water sample. $(X_{xx})_{distilled}$ is the imaginary part of the impedance value when the sensor is immersed in the distilled water/mili-q water and $(X_{xx})_{sample}$ is the imaginary part of the impedance value when the sensor is immersed in the water sample. The symbol *f* represents the frequency.

1.7 Light Sensing Technology

The measurement of light and light intensity is important for many applications, a few typical applications include auditorium, lecture theatre, library, forest canopies, greenhouses monitoring etc. In the presence of sunlight, the photosynthesis takes place where plants convert carbon dioxide and water into carbohydrates. Plants use light in the range of 400 to 700 nm. This range is most commonly referred to as PAR (Photo-synthetically Active Radiation). Monitoring PAR is important to ensure that plants are receiving adequate light for photosynthesis. A few light sensors available off-the-shelf which are used for environmental monitoring applications. Their advantages and disadvantages are also discussed.

1.7.1 Photometric Sensors

Photometric sensors are designed to measure visible radiation that has spectral response similar to that of human eye. Some of the applications for photometric sensor include interior and industrial lighting, outdoor illumination, illumination engineering and passive solar energy. Figure 1.19 shows a few photometric sensors commonly available in the market.

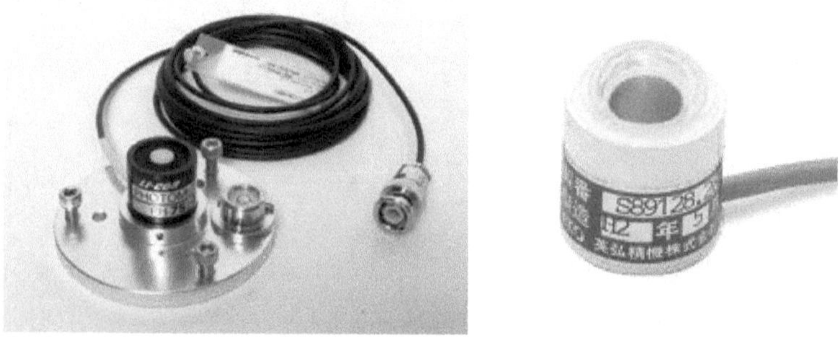

Fig. 1.19 Photographs of photometric sensors

The advantages of photometric sensors are: fast response time, highly sensitive, good stability, excellent linearity, low temperature dependency and small in size. But the sensors are expensive and they are mostly used to measure indoor lighting conditions. For environmental applications, PAR and Solar Radiation sensors are preferred.

1.7.2 Light Dependent Resistors (LDR)

The light dependent resistors (LDR) measure visible light as seen by the human eye. It is basically a resistor that has internal resistance which increases or

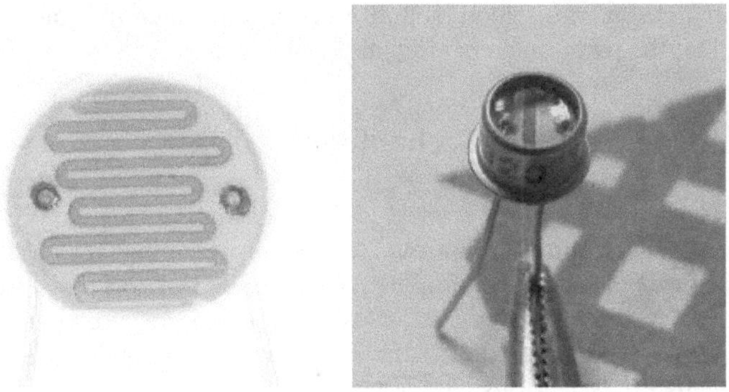

Fig. 1.20 Pictures of the light dependent resistors

decreases dependent on the level of light intensity impinging on the surface of the sensor. Figure 1.20 shows pictures of commonly available LDR sensors.

The advantages of LDRs are: comparatively cheap, linear output, fast response, and small in size. But the LDR sensors are mostly used to measure indoor lighting conditions.

1.7.3 Pyranometers

A pyranometer is an instrument used to measure the combined intensity of incoming direct solar radiation and diffuse sky radiation. The sensor is composed of a silicon photovoltaic detector mounted in a miniature head. It compares the heating produced by the radiation on blackened metal strips with that produced by an electric current. These sensors are commonly used for agriculture, meteorological and solar energy applications. Figure 1.21 shows a typical pyranometer system to measure the light intensity.

Fig. 1.21 Pyranometers

The advantages are: good stability, excellent linearity, fast response, and accurate. But the pyranometers are bulky and expensive.

1.8 Moisture Sensing Technology

Measurement of moisture content is important in a number of different applications; agricultural, houses, textiles, packaging materials, electronic appliances or dry food processing etc. Measurement of soil moisture is useful for minimizing the amount of irrigation water applied for growing plants and for optimizing plant growth. Due to the importance of knowledge of the moisture content of materials, various techniques have been developed to measure it. This section outlines a number soil moisture detection technologies are available on the market as well as addresses their advantages and disadvantages.

1.8.1 Frequency Domain Reflectometry (FDR) Soil Moisture Sensor

Frequency Domain Reflectometry is also referred to as a capacitance sensor. The sensor probes based on the FDR method of soil moisture measurement employ an oscillator to generate an electromagnetic signal that is propagated through the unit and into the soil. A part of this signal will be reflected back to the unit by the soil. This reflected wave is measured by the FDR probe to predict the water content of the soil. Figure 1.22 shows typical probe of FDR soil moisture sensor.

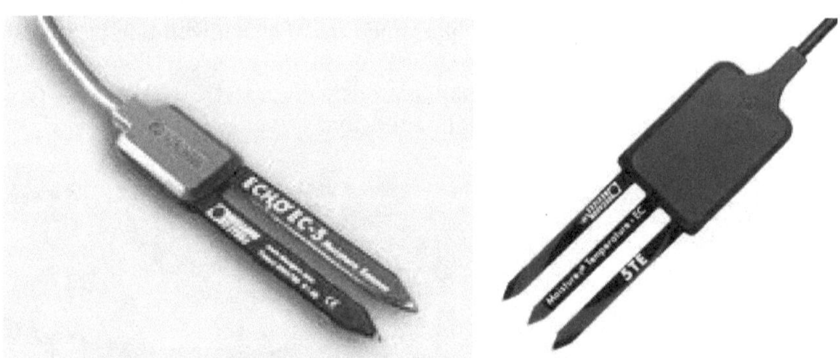

Fig. 1.22 Frequency Domain Reflectometry (FDR) Soil Moisture Sensors

The advantages are: fast response, accurate and inexpensive. But the disadvantage is that the sensor needs calibration with the soil the probes will be buried in.

1.8.2 Time Domain Reflectometry (TDR) Soil Moisture Sensor

Time Domain Reflectometry (TDR) sensors work by propagating a pulse down a line into the soil, which is terminated at the end by a probe with wave guides. TDR systems measure water content of the soil by measuring how long it takes the pulse to come back. Figure 1.23 shows typical probe of TDR soil moisture sensor.

The advantages are: fast response and accurate. But the disadvantage is that the sensor needs calibration which can be tedious and difficult, expensive and they are easy to corrode.

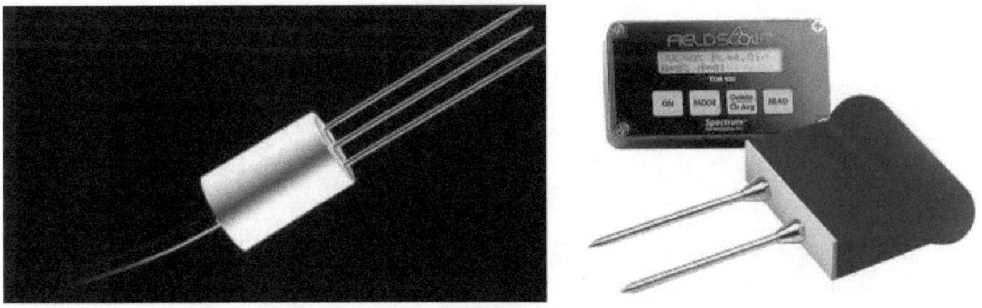

Fig. 1.23 Time Domain Reflectometry (TDR) Soil Moisture Sensor

1.8.3 Gypsum Blocks

Gypsum blocks use two electrodes placed into a small block of gypsum to measure the soil water tension. The amount of water in the soil is determined by the electrical resistance between the two electrodes within the gypsum block. More water present in the soil will reduce the resistance, while less water will increase it. Figure 1.24 shows a picture of gypsum blocks.

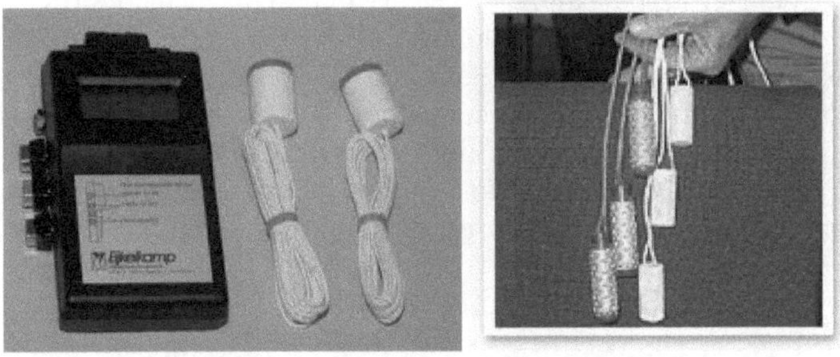

Fig. 1.24 Gypsum Blocks

The advantages are: inexpensive and easy to install. But the disadvantages are that they have to be replaced periodically and are sensitive to the saline content of the salt.

1.8.4 Neutron Probes

Neutron probes can be used to measure soil moisture content. A probe is inserted in the ground which emits low-level radiation in the form of neutrons. These neutrons collide with the hydrogen atoms contained in water, which is detected by the probe. The more water content in the soil, the more neutrons are scattered back at the device. Figure 1.25 shows the pictorial representation of the use of neutron probe to determine moisture in the soil.

Fig. 1.25 Neutron Probes

The advantages are: accurate and fast response. But the disadvantages are that they are expensive and the users have to be registered with the government due to radioactive elements used to emit the neutrons.

1.9 Carbon Dioxide (CO_2) Sensing Technology

The Measurement of carbon dioxide, CO_2 is important in monitoring indoor air quality and in many industrial processes. Carbon dioxide is one of the most common by-product of living organisms. The gas itself is safe in low concentration, however in high concentration it could be life threatening. Two types of CO_2 detectors are available to measure CO_2 level in the environment: Electrochemical CO_2 Sensors and Non-dispersive Infrared (NDIR) CO_2 Sensors. This section discusses their operating principles as well as addressing their advantages and disadvantages.

1.9.1 Solid State Electrochemical (SSE) CO_2 Sensors

Solid state electrochemical (SSE) CO_2 sensors are used in different applications which include greenhouse, farming, land fill gas monitoring, indoor air quality motoring and hazardous area warning signals. They adopt a galvanic structure with a sodium-ion conducting electrolyte of beta alumina, operating at 450°C. Carbonate based sensing material is deposited on one side as a sensing electrode and counter electrode is attached on the other side as a reference electrode. When the sensor is exposed to the environment, the sensor will generate voltage outputs in respect to the level of CO_2 in the environment. Figure 1.26 shows the picture of a off-the-shelf available electrochemical CO_2 sensor.

Fig. 1.26 Electrochemical CO_2 Sensors

The advantages are: relatively accurate, cheap, real-time sensing, high tolerance against contaminants and small in size. The disadvantages are: they require a significant amount of power because they operate at high temperature and need at least 24 hours of preconditioning before it could be used.

1.9.2 Non-dispersive Infrared (NDIR) CO_2 Sensors

Non-dispersive Infrared (NDIR) sensors are spectroscopic sensors that are capable of detecting CO_2 level in a gaseous environment based on its absorption characteristic. The key components are an infrared source, a light tube (or a chamber), an interference (wavelength) filter, and an infrared detector. The gas is pumped or diffuses into the light tube and the electronics measures the absorption of the characteristic wavelength of light. NDIR CO_2 sensor applications include indoor air quality monitoring, greenhouse farming, gas leak detection, automotive and flue gas emissions. Figure 1.27 shows the picture of a few off-the-shelf available Non-dispersive Infrared CO_2 sensor.

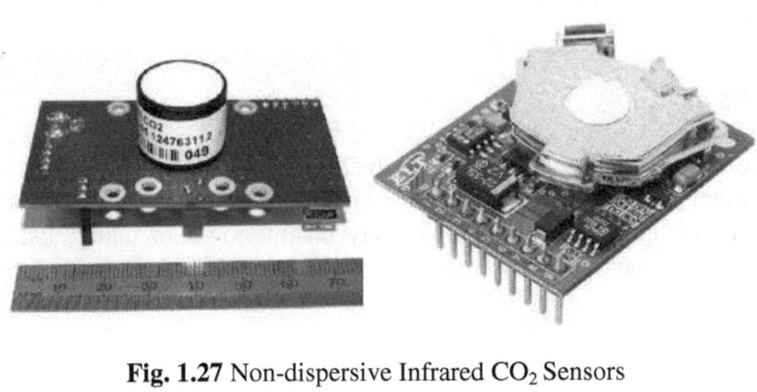

Fig. 1.27 Non-dispersive Infrared CO_2 Sensors

The advantages are: low power consumption, fast real-time sensing, high tolerance against contaminants and small in size. But the disadvantages are that carbon monoxide often coexists with CO_2 and absorbs a similar wave-length range as CO_2 which results in inaccurate measurement of CO_2 concentration and expensive.

1.10 Sensors Parameters

The sensors are the fundamental element for any monitoring, measurement and control system. There are different types of sensors available and they are to be properly chosen depending on the applications. For normal household applications, thermal sensors and humidity sensors are very commonly used. A few sensors such as strain gauges, accelerometers, temperature, acoustic emission sensors are very commonly used sensors used for monitoring the health of structures. Nowadays, the fiber optic based sensor systems are becoming very popular due to their different advantages compared to other sensors. The selection of sensors, cost, number of sensors and their placements, protection against mechanical and chemical damage, reduction of noise, and the collection of more representative data are the few things considered for sensors used for different applications. The sensitivity of sensors to moisture and humidity is another concern, especially when long-term measurement is planned, particularly in a harsh environment. Special provisions are often needed to protect the sensors in order to obtain acceptable measurements. Since the sensors are planned to be used for a long duration, the energy harvesting may also need to be considered. The sensors considered for health monitoring of structures are usually smart wireless sensors as wired sensors may not be a cheap and viable option for this type of applications. The sensors along with signal conditioning in combination with a microcontroller/microprocessor all come in the same package and can be defined as smart sensors or smart wireless sensors. Usually the smart sensors have the ability to compensate for random errors, can adapt to changes in the environment, can adjust non-linearities to give a linear output, have the provision of self-calibration and self-diagnosis of fault. The smart sensors have their own standard, IEEE 1451 so that they can be used in a ‘plug-and-play’ manner. The following characteristics are very important for the selection of sensors used for different applications.

1.10.1 Range

It is defined as the limits between which the inputs of the sensor can vary. It is very important for the sensors used for health monitoring of structures as the maximum input applied to the sensor may be unknown in many instances. The sensors should not be damaged at that abnormal condition.

1.10.2 *Sensitivity*

In general, the sensors used for any applications should be sensitive enough to give correct information on the effect of the input signal. The sensitivity is the relationship between output and input and is also used to indicate about the change of output to inputs other than being measured, such as environmental parameter changes. It is desirable that the sensitivity of the sensors to the environmental parameter changes is ideally zero or should be very small so that it can be easily neglected. If the effect of environmental parameters is not negligible, a compensation method may be introduced.

1.10.3 *Accuracy*

It is a measure of the closeness of the actual output to the ideal output of the sensor. It is an indication of the extent by which the measurement is wrong. It is the summation of all the possible errors that are likely to occur and it also depends on the calibration method. The accuracy may be represented either in absolute value or may be in percentage of the full range output.

1.10.4 *Stability*

Typically, the sensors used for any applications are in service continuously over many years. The sensors should be stable enough to give the same output for a constant input over a period of time. With respect to stability, a term 'drift' is used to describe the change in output that occurs over time. It is expressed as a percentage of the full range output.

1.10.5 *Repeatability*

The repeatability is very important for any sensors especially for the sensors used for any critical applications. It is the ability to give the same amount of output for repeated applications to the same amount of input. It is also termed as 'reproducibility'. The error is usually expressed as a percentage of the full range output:

$$\text{Repeatability}=\frac{\text{Maximum Output (for an input) - Minimum Output (for the same input)}}{\text{Full Range}}*100$$

It is expected that the repeatability of the sensors used for some critical applications such as Structural Health Monitoring (SHM) should be better than 0.01%.

1.10.6 Static and Dynamic Characteristics

While the sensors are selected for any applications, the static and dynamic characteristics of the sensors such as rise time, time constant and settling time should be looked into for selection. In some situation the slow response of the sensors may not be very critical for any applications such as the monitoring the health of structures. While the sensors are subjected to a dynamic input condition, the response should be free from hysteresis.

1.10.7 Energy Harvesting

The sensors employed for Wireless Sensor Networks (WSN) are used for many years. So it may be a good idea to investigate some kind of energy harvesting option so that the sensors will be self-sufficient in terms of operation.

1.10.8 Compensation Due to Change of Temperature and Other Environmental Parameters

The responses of the sensors are usually affected due to change of ambient temperature, humidity and other environmental parameters. To reduce the effect of the external influences, adequate compensation schemes must be included in the signal conditioning part of the sensors.

1.11 Selection of Sensors

The selection of sensors especially for students starting a project may be a difficult task and can waste a significant amount of time on searching internet. It is important to have a good idea on the type of sensors which are necessary for the desired application so that the specification of the sensor can be checked in the published datasheet. While a sensor is chosen for a specific application there are many criterions involved.

Operating Principle: It is very important to have a good idea of the different operating principles involved of the desired sensors.

Availability: It is very important to know the sources of the suppliers of the sensors, physical location, delivery schedule, payment options, continuation of supply etc. can have effect on the projects.

Cost: the actual cost of the sensor itself and the delivery cost involved may be very useful whether to order a few more number of sensors together.

Performance Figures: The performance figures such as (i) Range, (ii) Resolution, (iii) Repeatability, (iv) Accuracy, (v) Drift, (vi) Hysteresis effect, (vii) Ease of use and (viii) Power supply requirements need to be looked into.

Need of frequent calibration: Sometimes it may be important for many sensors to calibrate frequently.

Effect of other parameters: It is important to know how the environmental parameters such as change of temperature and humidity etc change the response of the sensors.

The above discussion may be used as a guidelines to select sensors.

1.12 Suggested Further Reading

The readers may read the following references for more knowledge and information.

[1] Mukhopadhyay, S.C., Sen Gupta, G.: Smart Sensors and Sensing Technology. LNEE. Springer (July 2008) ISSN 1876-1100, ISBN 978-3-540-79589-6

[2] Mukhopadhyay, S.C., Huang, R.Y.M.: Sensors, Advancements in Modeling, Design Issues, Fabrication and Practical Applications. LNEE. Springer (2008) ISSN 1876-1100, ISBN 978-3-540-69030-6

[3] Mukhopadhyay, S.C., Sen Gupta, G., Ray Huang, Y.M.: Recent Advances in Sensing Technology. LNEE. Springer (September 2009) ISSN 1876-1100, ISBN 978-3-642-00577-00

[4] Mukhopadhyay, S.C., Lay-Ekuakille, A.: Advances in Biomedical Sensing, Measurements, Instrumentation and Systems. LNEE. Springer (January 2010) ISSN 1876-1100, ISBN 978-3-642-05166-1

[5] Mukhopadhyay, S.C., Leung, H.: Advances in Wireless Sensors and Sensors Networks. LNEE. Springer (July 2010) ISSN 1876-1100, ISBN 978-3-642-12706-9

[6] Lay-Ekuakille, A., Mukhopadhyay, S.C.: Wearable and Autonomous Biomedical Devices and Systems for Smart Environment: Issues and Characterization. LNEE. Springer (September 2010) ISSN 1876-1100, ISBN 978-3-642-15686-1

[7] Mukhopadhyay, S.C., Lay-Ekuakille, A., Fuchs, A.: New Developments and Applications in Sensing Technology. LNEE, vol. 83. Springer (January 2011) ISBN 978-3-642-17942-6

[8] Mukhopadhyay, S.C.: New Developments in Sensing Technology for Structural Health Monitoring. LNEE, vol. 96. Springer (August 2011) ISBN 978-3-642-21098-3

[9] Mukhopadhyay, S.C.: Smart Sensing Technology for Agricultural and Environmental Monitoring. LNEE, vol. 146. Springer (February 2012) ISBN 978-3-642-27637-8

[10] Suryadevara, N.K., Gaddam, A., Rayudu, R.K., Mukhopadhyay, S.C.: Wireless Sensors Network based safe Home to care Elderly People: Behaviour Detection. Sens. Actuators A: Phys. (2012), doi:10.1016/j.sna.2012.03.020

[11] Amri Yunus, M., Mukhopadhyay, S.C., Ibrahim, S.: Planar Electromagnetic Sensors Based Estimation of Nitrates Contamination in Water Sources Using Independent Component Analysis. IEEE Sensors Journal 12(6), 2024–(2033)

[12] Rahman, M.S.A., Mukhopadhyay, S.C., Yu, P.L., Chuang, C.H., Haji-Sheikh, M.: Measurements and Performance Evaluation of Novel Interdigital Sensors for Different Chemicals Related to Food Poisoning. IEEE Sensors Journal 11(11), 2957–2965 (2011)

[13] Ranhotigamage, C., Mukhopadhyay, S.C.: Field Trials and Performance Monitoring of Distributed Solar Panels Using a Low Cost Wireless Sensors Network for Domestic Applications. IEEE Sensors Journal 11(10), 2583–2590 (2011)

[14] Amri Yunus, M., Mukhopadhyay, S.C.: Novel Planar Electromagnetic Sensors for Detection of Nitrates and Contamination in Natural Water Sources. IEEE Sensors Journal 11(6), 1440–1447 (2011)

[15] Kaur, K., Mukhopadhyay, S.C., Schnepper, J., Haefke, M., Ewald, H.: A Zigbee Based Wearable Physiological Parameters Monitoring System. IEEE Sensors Journal 12(3), 423–430 (2012)

[16] Gaddam, A., Mukhopadhyay, S.C., Sen Gupta, G.: Elderly Care Based on Cognitive Sensor Network. IEEE Sensors Journal 11(3), 574–581 (2011)

[17] Mohd Syaifudin, A.R., Mukhopadhyay, S.C., Yu, P.L.: Novel Sensors for Food Inspection. Sensors and Transducers Journal 108, 1–44 (2010) ISSN 1726-5479

[18] Mohd Syaifudin, A.R., Jayasundera, K.P., Mukhopadhyay, S.C.: A Low Cost Novel Sensing System for Detection of Dangerous Marine Biotoxins in Seafood. Sensors and Actuators B: Chemical 137, 67–75 (2009), doi:10.1016/j.snb.2008.12.053

[19] Bhardwaj, S., Lee, D.S., Mukhopadhyay, S.C., Chung, W.Y.: Ubiquitous Healthcare Data Analysis And Monitoring Using Multiple Wireless Sensors for Elderly Person. Special Issue on Modern Sensing Technologies, Sensors and Transducers Journal 90, 87–99 (2008)

[20] Mukhopadhyay, S.C., Deb Choudhury, S., Allsop, T., Kasturi, V., Norris, G.E.: Assessment of pelt quality in leather making using a novel non-invasive sensing approach. Journal of Biochemical and Biophysical methods, JBBM 70, 809–815 (2008)

[21] Reichard, J., Brown, C.: Detecting groundwater contamination of a river in Georgia, USA using baseflow sampling. Hydrogeology Journal 17, 735–747 (2009)

[22] Maria Dolores, F.R., et al.: The use of one-shot sensors with a dedicated portable electronic radiometer for nitrate measurements in aqueous solutions. Measurement Science and Technology 19, 095204 (2008)

[23] Mesina, M.B., de Jong, T.P.R., Dalmijn, W.L.: Automatic sorting of scrap metals with a combined electromagnetic and dual energy X-ray transmission sensor. International Journal of Mineral Processing 82, 222–232 (2007)

[24] Guadarrama-Santana, A., Garcia-Valenzuela, A.: Principles and Methodology for the Simultaneous Determination of Thickness and Dielectric Constant of Coatings With Capacitance Measurements. IEEE Transactions on Instrumentation and Measurement 56, 107–112 (2007)

[25] Kirchner, N., Hordern, D., Liu, D.K., Dissanayake, G.: Capacitive sensor for object ranging and material type identification. Sensors and Actuators a-Physical 148, 96–104 (2008)

[26] Sundara-Rajan, K., Byrd, L., Mamishev, A.V.: Moisture content estimation in paper pulp using fringing field impedance Spectroscopy. IEEE Sensors Journal 4, 378–383 (2004)

[27] Radke, S.M., Alocilja, E.C.: Design and fabrication of a microimpedance biosensor for bacterial detection. IEEE Sensors Journal 4, 434–440 (2004)

[28] Tan, E.L., Ng, W.N., Shao, R., Pereles, B.D., Ong, K.G.: A wireless, passive sensor for quantifying packaged food quality. Sensors 7, 1747–1756 (2007)

[29] Camurri, A., Lagerlof, I., Volpe, G.: Recognizing emotion from dance movement: comparison of spectator recognition and automated techniques. International Journal of Human-Computer Studies 59, 213–225 (2003)

[30] Healey, J.A., Picard, R.W.: Detecting stress during real-world driving tasks using physiological sensors. IEEE Transactions on Intelligent Transportation Systems 6(2), 156–166 (2005)

[31] Sierra, A.S., Avíla, C.S., Casanova, J.G., Pozo, G.B.: A stress detection system based on physiological signals and fuzzy logic. IEEE Transactions on Industrial Electronics, 4857–4865 (October 2011)

[32] Octavian, A.P., Pedro, M.B., Silva, G., Joaquim, M., Eduardo, C.P., Gabriela, P.: Physiological Parameters Measurement Based on Wheelchair Embedded Sensors and Advanced Signal Processing. IEEE Transactions on Instrumentation and Measurement 59(10), 2564–2574 (2010)

[33] Mamishev, A.V., Sundara-Rajan, K., Yang, F., Du, Y.Q., Zahn, M.: Interdigital sensors and transducers. Proceedings of the IEEE 92, 808–845 (2004)

Chapter 2
Interfacing of Sensors and Signal Conditioning

Introduction

In this chapter, a few signal conditioning methods to interface analog and digital signals to processors, microcontroller, microprocessors etc will be introduced. The readers will get a good idea about different stages required to make an intelligent sensing and measurement system. The process involves passive and/or active devices and use circuits such as filters, current-to-voltage converters or vice-versa and so on. The main purpose of this chapter is to provide some fundamentals of interfacing the signals obtained from different sensors to be available to the digital processor for the necessary computation. The digital signal may be directly connected to a digital input of a processor provided it is within the required voltage range. The analog signals are usually connected to the Analog-to-Digital converters of the processor.

2.1 Change of Bias and Level of Signals

In most applications the sensor signals need some form of change of bias and/or shift of voltage level. For example, the output voltage available from the sensor may vary from 0.2 V to 1.3 V as a change of the input variable over the entire range of measurement. When the output of the sensor is connected to a microcontroller, it may require a voltage level from 0 V to 3.3 V for the same variation of the input variable.

The required signal conditioning is done by changing the output of the sensor from 0.2 V to 0 V with the input variable is 0 V. This may be done simply by subtracting 0.2 V from the output of the sensor. This may be defined as a bias adjustment or zero shift of the output.

Once the bias adjustment is done, the output of the sensor is now varying from 0 V to 1.1 V. In order to utilise the full range of 0 V to 3.3 V of the processor, the signal is multiplied by a gain of 3. This process is known as amplification and the amplifying factor (3 in this case) is known as gain. In many situations, the output to the processor need to be reduced, this is known as attenuation. Both the amplification and attenuation is achieved by the same circuit, known as amplifier. The gain of the amplifier is more than unity for an amplifier and is less than unity for an attenuator.

S.C. Mukhopadhyay: *Intelligent Sensing, Instrumentation and Measurements*, SSMI 5, pp. 29–53.
DOI: 10.1007/978-3-642-37027-4_2

It is important to take note of the frequency response, input impedance and output impedance while the bias and amplifier circuits are designed.

2.2 Loading Effect on Sensor's Output

It is useful to understand the loading effect of sensor's output while designing the signal conditioning circuit. The voltage measured across the output of the sensor is the open-circuit voltage. When the sensor is connected to a circuit, the voltage across the terminals of the sensor drops down to a value and is less than the open-circuit voltage. This is explained with the help of figure 2.1. The output of the sensor is represented by the source voltage V_s. The impedance of the sensor (the output resistance) is R_s. The sensor is connected to a load of resistance R_{load} which can be the input resistance of any amplifier or any port pin of the processor.

Under open-circuit condition the output of the sensor is V_s.

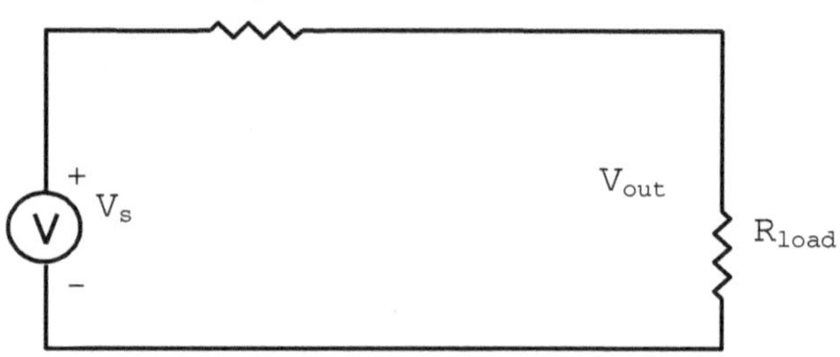

Fig. 2.1 The connection of sensor to explain loading effect

From the figure 2.1, the output V_{out} across the resistance R_{load}, can be calculated as

$$V_{out} = \frac{R_{load}}{R_s + R_{load}} V_s \tag{2.1}$$

$$\frac{V_{out}}{V_s} = \frac{1}{1 + \frac{R_s}{R_{load}}} \tag{2.2}$$

From (2.2), it can be said that the output voltage V_{out} will be equal to V_s only when the load resistance R_{load} is much larger that the source resistance R_s. The effect of loading can be illustrated with the help of the following example.

Ex 2.1: A temperature sensor provides an output of 1 mV/°C and has an output resistance of 1 kΩ. The sensor is connected to an amplifier of input resistance of 4 kΩ. If the gain of the amplifier is 100, find the output of the amplifier for a temperature of 50°C.

Sol: The simple solution will be as follows:

The output of the sensor at 50°C, is 50 * 1 mV = 50 mV.

The input to the amplifier is 50 mV and the gain is 100. So the output of the amplifier is 100 * 50 mV = 5000 mV = 5 V.

But, the above result is wrong. The actual situation is shown in figure 2.2.

The input to the amplifier is given by

$$V_{in} = \frac{R_2}{R_1+R_2} * V_s = \frac{4}{1+4} * 50\text{ mV} = 40\text{ mV}.$$

So, the actual output of the amplifier, Vout = 100* 40 mV = 4000 mV = 4 V.

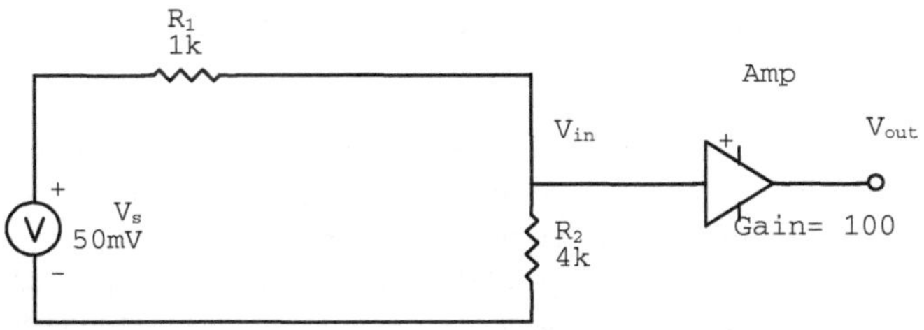

Fig. 2.2 The representation of the connection of the temperature sensor

2.3 Potential Divider

The potential divider circuit is a common technique used for obtaining a reduced voltage. With the proper selection of the resistances any voltage can be obtained which is lower than the input voltage.

The output voltage is obtained with the help of the potential divider formed by the resistances R_1 and R_2 as shown in figure 2.3 and is given by in (2.3):

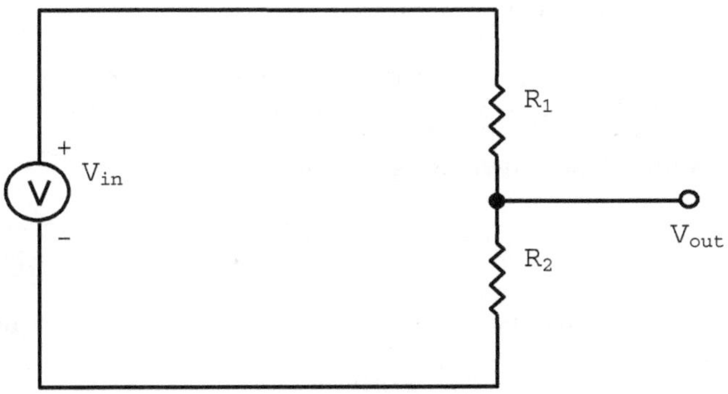

Fig. 2.3 A simple potential divider scheme

$$V_{out} = \frac{R_2}{R_1 + R_2} * V_{in} \tag{2.3}$$

Where R_1 and R_2 are the divider resistances, V_{in} is the input voltage and V_{out} is the desired voltage at the output.

Either of the resistances R_1 and R_2 used in the potential divider may represent a sensor whose resistance varies as a function of the measured variable. A few issues need to consider while the resistance values for the potential divider are chosen:

i. A current flows continuously through both the resistors and a power will be dissipated. The power dissipation and consequently the power rating of the resistors need to be considered.
ii. The output voltage doesn't vary linearly with the divider resistances.
iii. The loading effect should be considered. The potential divider may be considered as a voltage source, given by (2.3) in series with a Thevenin's resistance. The Thevenin's resistance is the parallel combination of the two resistances R_1 and R_2. The equivalent Thevenin's resistance will be less than the minimum value of the resistances R_1 and R_2. When the output V_{out} is connected to an amplifier or to any other circuit, the input resistance of the amplifier should be very large compared to the Thevenin's resistance of the divider circuit.

Ex 2.2: In the figure 2.3, the input voltage V_{in} is 1 V. The resistance R_2 is a fixed resistance of 10 kΩ and the resistance R_1 represents a sensor whose resistance varies from 5 kΩ to 15 kΩ for the whole measurement range. Determine the maximum and minimum value of the output voltages and corresponding equivalent Thevenin's resistances. How much is the power dissipated in the sensor?

Sol: Following the (2.3), the output voltage can be calculated as

$$V_{out} = \frac{R_2}{R_1 + R_2} * V_{in};$$

The maximum value of the output voltage in volt,

$$V_{out,Max} = \frac{R_2}{R_1 + R_2} * V_{in} = \frac{10}{5+10} * 1 = 0.666$$

The minimum value of the output voltage in volt,

$$V_{out,Min} = \frac{R_2}{R_1 + R_2} * V_{in} = \frac{10}{15+10} * 1 = 0.4$$

The corresponding values of the Thevenin's resistances are obtained using

$$R_{Th} = \frac{R_1 R_2}{R_1 + R_2}$$

So, the Thevenin's resistance corresponding to the maximum voltage situation,

$$R_{Th} = \frac{5*10}{5+10} = 3.33\ \text{k}\Omega$$

The Thevenin's resistance corresponding to the minimum voltage situation,

$$R_{Th} = \frac{15*10}{15+10} = 6 \text{ k}\Omega.$$

The power dissipation in the sensor can be calculated as follows:

During the maximum output voltage condition, the voltage across the resistance R_l is (1 – 0.666) V = 0.333 V.

So the power dissipated at R_l is = $(0.333)^2/5$ mW = 0.0221 mW.

During the minimum output voltage condition, the voltage across the resistance R_l is (1 – 0.4) V = 0.6 V.

So the power dissipated at R_l is = $(0.6)^2/15$ mW = 0.024 mW.

2.4 Low-Pass RC Filter

It is often necessary to eliminate unwanted noise signals in the sensor's output before it is interfaced to processors. This is achieved by using a simple RC circuit, known as filter. A simple low-pass RC filter is constructed from a single resistor and a capacitor as is shown in figure 2.4.

The output expression can be written as

$$\frac{V_{out}}{V_{in}} = \frac{\frac{1}{j\omega C}}{R+\frac{1}{j\omega C}} = \frac{1}{1+j\omega RC} = \frac{1}{1+j\frac{\omega}{\omega_C}} \tag{2.4}$$

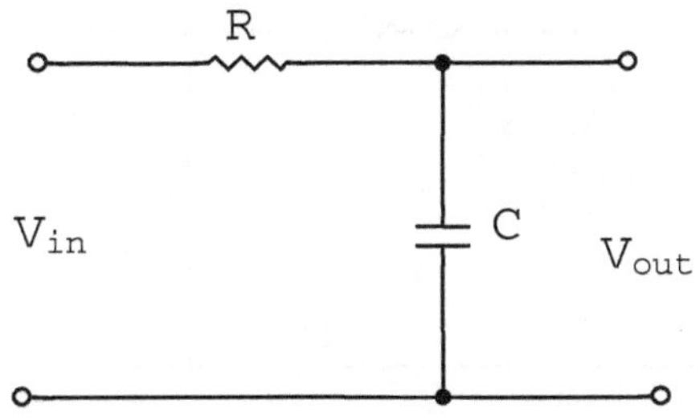

Fig. 2.4 A simple low-pass RC filter

$$\text{Where } \omega_c = 1/(\text{RC}) \tag{2.5}$$

and is known as angular cut-off frequency. The signal passes to the output terminal without any attenuation for any frequency below the cut-off frequency. For any frequency above ω_c the signal gets attenuated.

The absolute output-to-input voltage ratio for any signal frequency can be written as

$$\left|\frac{Vout}{Vin}\right| = \frac{1}{[1+(f/f_C)^2]^{\frac{1}{2}}} \tag{2.6}$$

$$\text{Where } f_c = \omega_c/(2\pi) = \frac{1}{2\pi RC} \tag{2.7}$$

f_c is the cut-off frequency in Hertz (Hz). It is seen from (2.6) that at $f = f_c$, the ratio of the output to input voltage is $1/\sqrt{2} = 0.707$.

The students face the situations of designing a simple RC filter very frequently with a known critical frequency. The task is to find the suitable values of R and C to accomplish the filtering performance. The students need to select two components R and C from one equation (2.7). By following the guidelines, they can find the values of R and C.

1. First, a standard capacitor in the range of pF to μF is selected.
2. Using (2.7), the resistor value is calculated. It is better that the value of the resistance is within kΩ and MΩ range. This is to avoid the loading and noise problem. If the range is outside the kΩ and MΩ range, it is better to change the capacitor value to get the resistance value with the kΩ and MΩ range.
3. The standard value of resistance can be chosen which is close to the calculated value.
4. If the design allows some flexibility, the cut-off frequency may be slightly shifted for the chosen values of R and C. If the cut-off frequency is to be strictly maintained, the calculated value of resistance may be obtained by adding a potentiometer in series with a fixed resistance as is shown in figure 2.5.
5. The tolerance value of the capacitor and resistor should be taken into consideration.

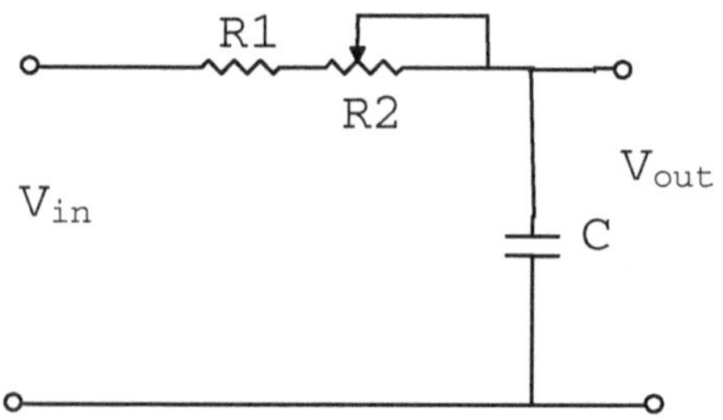

Fig. 2.5 A simple low-pass RC filter to achieve the desired cut-off frequency

The following design example will make it clear.

Ex 2.3: A low-pass RC filter is to be designed to attenuate the undesired noise signal. The range of the useful signal is up to 1 kHz. The filter should reduce the noise to 1% at 100 kHz.

Sol: From (2.6), we get

$$0.01 = \frac{1}{[1+(f/f_c)^2]^{\frac{1}{2}}}$$

We get, $1 + (\frac{f}{f_c})^2 = 10000$; Or, $(\frac{f}{f_c})^2 = 9999$

$As, f = 100$ kHz we get, $f_c = 1$ kHz.

By choosing a capacitor, C of 0.1 μF, we can calculate the value of the resistor, R from (2.7) as

$$R = \frac{1}{2\pi C f_c} = \frac{1}{2\pi * 0.1 * 10^{-6} * 1000} = 1.59\ \text{k}\Omega.$$

A standard value of 1.5 kΩ can be chosen. That will make the cut-off frequency as

$$f_c = \frac{1}{2\pi RC} = \frac{1}{2\pi * 1.5 * 10^3 * 0.1 * 10^{-6}} = 1061\ \text{Hz}.$$

With the cut-off frequency of 1061 Hz, the noise level at 100 kHz will be

$$\left|\frac{V_{out}}{V_{in}}\right| = \frac{1}{[1+(f/f_c)^2]^{\frac{1}{2}}} = \frac{1}{[1+(100*10^3/1061)^2]^{\frac{1}{2}}} = 0.0106.$$

So, a R of 1.5 kΩ and a C of 0.1 μF will do the desired objective.

2.5 High-Pass RC Filter

Sometimes, the requirement may be opposite to a low-pass filter, we may need to block low-frequency signal but allow the high-frequency signals to the output. This type of filter is known as high-pass filter. The simple RC high-pass filter is constructed by swapping the position of the resistor and the capacitor as is shown in figure 2.6.

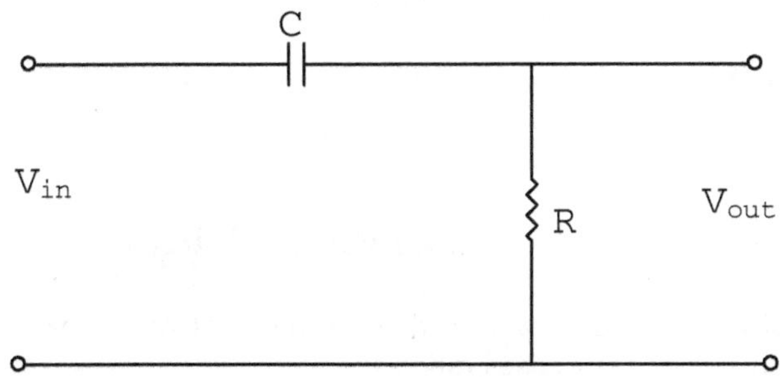

Fig. 2.6 A simple high-pass RC filter

The output expression can be written as

$$\frac{V_{out}}{V_{in}} = \frac{R}{R+\frac{1}{j\omega C}} = \frac{j\omega RC}{1+j\omega RC} = \frac{j\frac{\omega}{\omega_c}}{1+j\frac{\omega}{\omega_c}} \qquad (2.8)$$

Where $\omega_c = 1/(RC)$ and is known as angular cut-off frequency. The signal passes to the output terminal without any attenuation for any frequency above the cut-off frequency. For any frequency below ω_c the signal gets attenuated. When the frequency is equal to ω_c, the critical frequency, the ratio of V_{out}/V_{in} is 0.707.

The absolute output-to-input voltage ratio for any signal frequency can be written as

$$\left|\frac{Vout}{Vin}\right| = \frac{\langle f/fc\rangle}{[1+\langle f/fc\rangle^2]^{\frac{1}{2}}} \tag{2.9}$$

Where $f_c = \omega_c/(2\pi) = \frac{1}{2\pi RC}$.

The following design example will make it clear.

Ex 2.4 A high-pass RC filter is to be designed to attenuate the undesired 50 Hz supply frequency noise. The useful signal is 1 kHz and above. Design the suitable R and C of the filter so that the 50 Hz signal gets eliminated and the useful signal is down only by 3 dB.

Sol: In terms of dB, we can write dB = 20 log (V_{out}/V_{in}).

Down by 3 dB, means 20 log (V_{out}/V_{in}) = -3 which means

V_{out}/V_{in} = 10^(-3/20) = 0.707.

From (6.9), we have $V_{out}/V_{in} = 0.707$ at $f = f_c$.

So, the critical frequency of the filter is f_c = 1 kHz.

Let us first check the magnitude of the 50 Hz signal at 1 kHz.

From (6.9), we have $\left|\frac{V_{out}}{V_{in}}\right| = \frac{\langle f/f_c\rangle}{[1+\langle f/f_c\rangle^2]^{\frac{1}{2}}} = \frac{\langle 50/1000\rangle}{[1+\langle 50/1000\rangle^2]^{\frac{1}{2}}} = \frac{0.05}{1.001249} = 0.0499$.

So, around 5% of 50 Hz signal is present at 1 kHz. This means that the 50 Hz supply noise has been reduced by 95%.

For designing the R and C components, we can choose a capacitor, C of 0.01 μF. We can calculate the value of the resistor, R from (2.7) as

$$R = \frac{1}{2\pi f_c C} = \frac{1}{2\pi * 1000 * 0.01 * 10\text{^}{-6}} = 15.9\ k\Omega.$$

A standard value of 15 kΩ can be chosen. That will make the cut-off frequency as $f_c = \frac{1}{2\pi RC} = \frac{1}{2\pi * 15 * 10^3 * 0.01 * 10\text{^}{-6}} = 1061$ Hz.

2.6 Practical Issues of Designing Passive Filters

A few practical issues need to be considered while passive low-pass or high-pass RC filters are designed: They are as follows:

1. The product value of RC is given by (2.7). So a very small value of R and a very large value of C or vice-versa can satisfy the equation. But for practical consideration, a very small value of R should be avoided from loading consideration. In a similar way, a large capacitance value should also be avoided. As a thumb rule, it is better to keep the resistance in kΩ and above and the capacitance in μF and less.
2. Due to the use of filter, the effective input and output impedance of the circuit gets changed. Sometimes, a voltage follower may be added at suitable place to avoid the problem.
3. In many situations the exact critical frequency may not be so important. So the critical frequency achieved by the fixed value of resistance and

capacitance may be good enough. Though the exact value of critical frequency may be achieved by using potentiometer in series with a fixed resistance but it increases cost.

4. The use of filter increases the order of the system equation.

2.7 Op-Amp Based Instrumentation

Operational amplifier (op-amp) is very commonly used in many sensor applications to solve interfacing problems. The description of op-amp, their characteristics and different schematics are available in any standard book. So we will not spend time on that. In this book we will describe some schemes which are commonly used in sensors applications.

We will start with a simple problem.

Ex 2.5 A current sensor of very weak power source provides 10 mV for its full range. The sensor is to be interfaced to a processor of maximum range of 1.2 V. Design a suitable op-amp based circuit with schematic diagram.

Sol: A high input impedance non-inverting amplifier is required for this objective. The op-amp based schematic configuration is shown in figure 2.7. The sensor is directly connected to the non-inverting input of the op-amp. The input impedance of this scheme is very high as it is effectively the input impedance of the op-amp. The output impedance is very low.

The gain = 1.2 V/ 10 mV = 120.

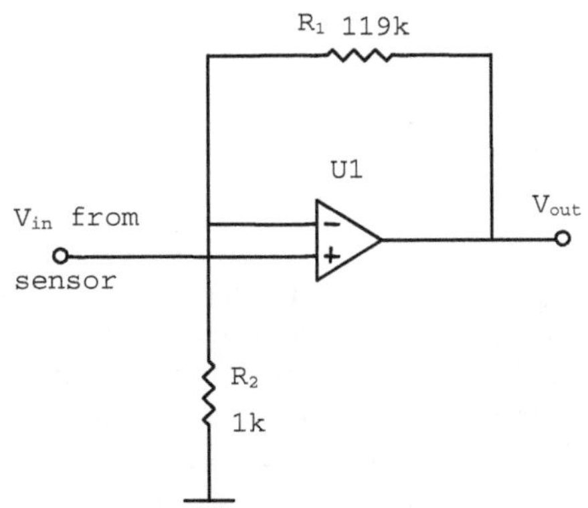

Fig. 2.7 A simple high-impedance amplifier with a gain of 120

The output expression is given by,

$$\frac{V_{out}}{V_{in}} = 1 + \frac{R_1}{R_2} \tag{2.10}$$

This gives $R_1/R_2 = 119$.

By choosing, $R_2 = 1$ kΩ, we can have $R_1 = 119$ kΩ. Since 120 kΩ is a standard resistance, R_1 can be chosen as 120 kΩ.

2.7.1 Differential Amplifier

The circuit as is shown in figure 2.8 is a very simple scheme to achieve the output signal which is an amplified version of the difference of the two input signals. The scheme is based on two pairs of matching resistances R_1 and R_2. Based on the ideal matching resistances, the output equation is given by

$$V_{out} = \frac{R_1}{R_2}(V_1 - V_2) \tag{2.11}$$

where V_1 and V_2 are the voltages of the two inputs with respect to ground.

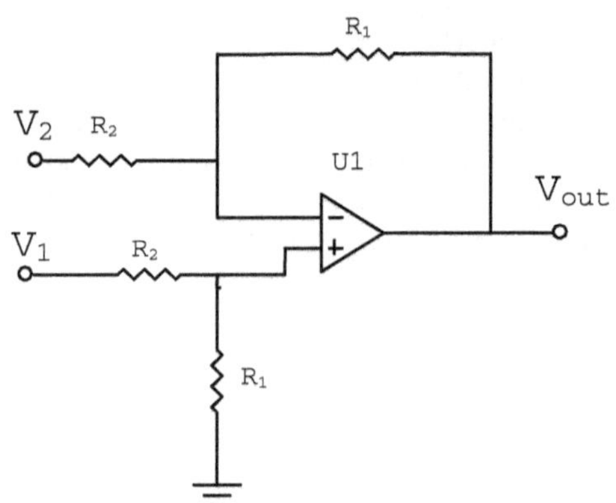

Fig. 2.8 A differential amplifier

The ratio R_1/R_2 provides the gain of the amplifier. If the resistances are not properly matched the common mode rejection will be poor. One disadvantage of the circuit is the low input impedance and the input impedances are not the same for both the inputs. In order to avoid this problem, voltage followers can be used on the input to provide high input impedance. The scheme is shown in figure 2.9 which is commonly known as instrumentation amplifier. The instrumentation amplifiers are characterized with a very high input impedance and a very low output impedance.

Ex 2.6: A temperature sensor provides an output of 10 mV to 210 mV over its operating range. Develop a signal conditioning circuit so that the output becomes 0 V to 3.3 V for the operating range. The circuit should offer very high input impedance.

Sol: Though there is only one input, i.e., only from the sensor, the problem can be configured by using an instrumentation amplifier. The other input can be provided from a source to nullify the effect of the offset voltage of the sensor.

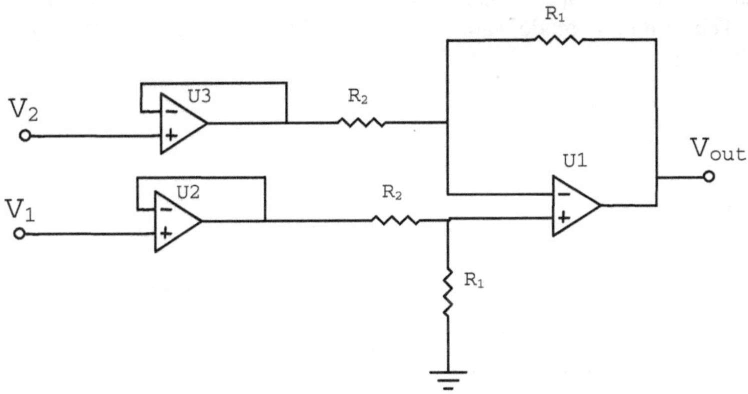

Fig. 2.9 An instrumentation amplifier

The schematic representation is shown in figure 2.10. The gain of the amplifier is 3.3 V/(201-10) mV = 16.5.

From the figure 2.10, it is seen that a voltage of 0.01 V (10 mV) has been produced from a source. A silicon diode has been used to make a source of 0.7 V. From which a voltage of 10 mV has been derived using a potential divider. For more stable voltage a zener diode can be used replacing the silicon diode.

2.7.2 *Common Mode Rejection*

The differential amplifier amplifies the difference of the input voltages, it does not depend on the values or the polarity of the individual input voltages. In order to quantify the performance of the differential amplifier Common Mode Rejection Ratio (CMRR) is used. The CMRR is defined as the ratio of the differential gain to the common mode gain.

If the input voltages are V_1 and V_2, the common mode voltage, $V_{com} = (V_1 + V_2)/2$ and the differential mode voltage, $V_d = (V_1 - V_2)$.

Assuming the gain of the amplifiers are, A_d is the differential gain and A_{com} is the common mode gain.

$$\text{So, the CMRR} = \frac{A_d}{A_{com}} \tag{2.12}$$

When the CMRR is expressed in dB, it is defined as the Common Mode Rejection, CMR.

$$\text{So, the CMR} = 20 \log(\text{CMRR}) \tag{2.13}$$

An ideal differential amplifier should have a very large CMR, typically more than 60 dB. The common mode gain A_{com} should be zero and the differential gain should be large.

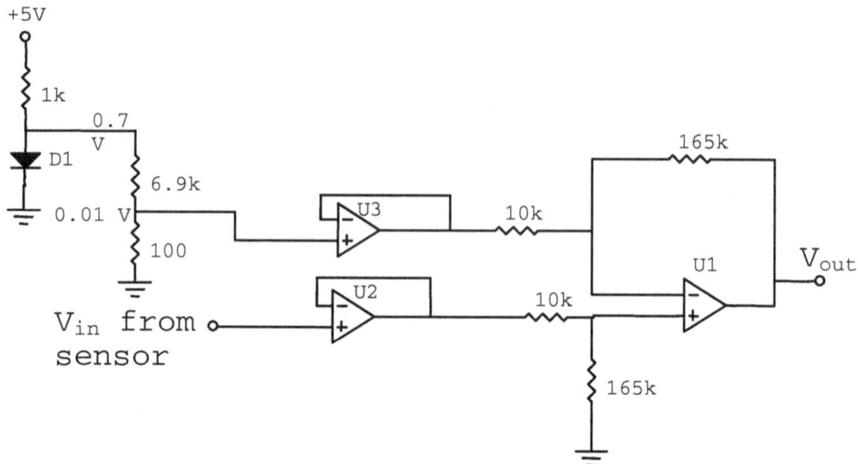

Fig. 2.10 The schematic representation of the solution of Ex 2.6

Ex 2.7 In a difference amplifier, the differential gain, A_d is 100. When an input of 1.5 V is fed to each input terminal, the output is 2 mV. Find the CMRR and CMR.

Sol: The common mode voltage, V_{com} = (1.5 + 1.5) / 2 = 1.5 V.

The common mode gain, A_{com} = 2 mV/ 1.5 V = 20*10^-3/15.

Then, the CMRR = A_d/A_{com} = 100/(20*10^-3/15) = 100*15*1000/20 = 75000.

And the CMR = 20 log CMRR = 20 log(75000) = 20 * 4.875 = 97.5 dB.

2.7.3 *Single-Resistance Controlled Instrumentation Amplifier*

Ideally, the common-mode gain of an instrumentation amplifier is zero. In the circuits shown in figures 2.8 and 2.9, common-mode gain is caused by mismatches in the values of the equally-numbered resistors and by the mis-match in common mode gains of the two input op-amps. It is difficult to obtain very closely matched resistors in practice. Though it is possible to build instrumentation amplifiers with individual op-amps and precision resistors, but are also available in integrated circuit form from several manufacturers (including Texas Instruments, National Semiconductor, Analog Devices, Linear Technology and Maxim Integrated Products). An Integrated Circuit based instrumentation amplifier typically contains closely matched laser-trimmed resistors, and therefore offers excellent common-mode rejection. Typical examples include AD620, MAX4194, LT1167 and INA128. It is possible to control the gain by adjusting the value of one resistor, R_{ext} as shown in figure 2.11.

The relationship of the output voltage and input voltages is given by the following equation:

$$V_{out} = \left(1 + \frac{2R_3}{R_{ext}}\right)\left(\frac{R_1}{R_2}\right)(V_1 - V_2) \tag{2.14}$$

The resistances, R_1, R_2 and R_3 are within the integrated circuit. The differential amplifier will have a fixed gain from the manufacturer. But the overall gain of the instrumentation amplifier can be changed by changing the R_{ext}. So, the high CMR can be obtained.

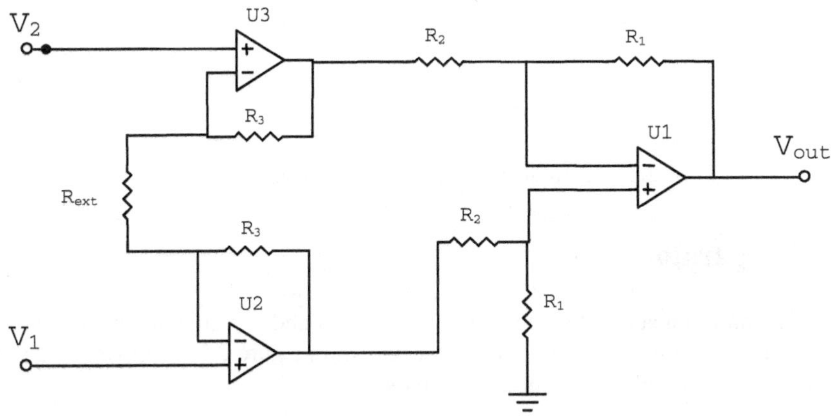

Fig. 2.11 Integrated circuit based instrumentation amplifier with one resistor for controlling the gain

2.8 Current-to-Voltage Converter

In many process control applications the signals are transmitted through long wires and are usually in the form of current of 4 mA to 20 mA. It is often necessary to convert the current into voltage and is done using a current-to-voltage converter circuit as is shown in figure 2.12. For many wired communication, the current is used to transmit measurement data of the controlled variable to the receiving end. Current is preferred over voltage as the system becomes less dependent on load and line resistance. As long as the current is held constant, the output will be independent to external condition and influences.

The relationship of the output voltage and the input current is given by,

$$V_{out} = - IR \tag{2.15}$$

A few things must be kept in mind while the current-to-voltage converter is designed. The current should not exceed the current carrying capability of the operational amplifier. The output voltage should not reach to the saturation voltage of the amplifier. Another resistance of R may be connected to the non-inverting terminal of the amplifier to provide temperature stability to the scheme.

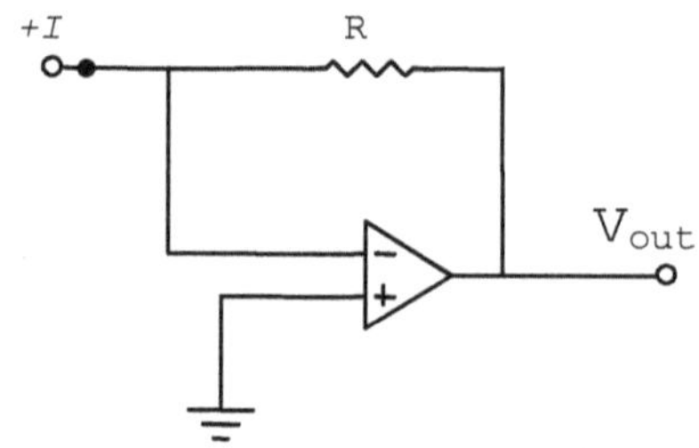

Fig. 2.12 An op-amp based current-to-voltage converter

2.9 Comparator

The basic communication between an analog signal and a digital signal with a 1/0 answer is a comparator. A comparator as is shown in figure 2.13 compares the two analog signals available on its input terminals.

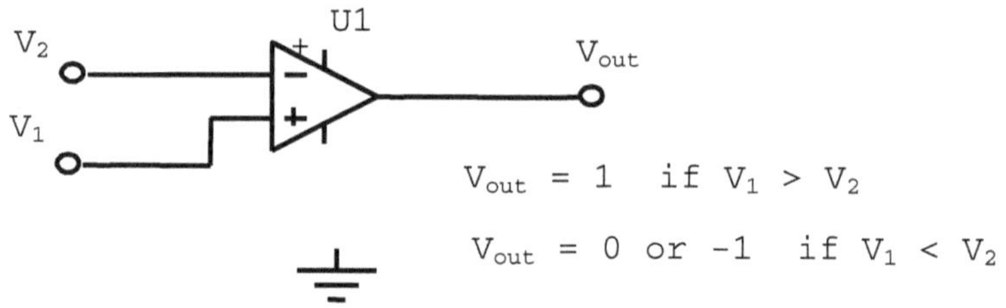

Fig. 2.13 The schematic representation of a basic comparator

The comparator will give a HIgh (1) output if the input V_1 (the non-inverting input) is larger than the input V_2 (inverting input). The comparator will provide a LOw (either 0 or -1) if the input V_2 is larger than input V_1. Usually one of the inputs is the reference input and is fixed. The reference voltage is usually derived from supply potentials and is dependent on the specification of the problem. It is assumed that the reference input will be maintained constant throughout the operation. The other input is variable in nature.

Usually the output of the comparator is the collector of an output transistor of the integrated circuit (IC). The collector is not connected to any power supply and is kept open. This configuration is called open-collector output configuration. So, it should be connected to a supply through a resistance before any output is observed or measured. Figure 2.14 shows the schematic representation of the comparator with the connection of a pull-up resistor to derive its output.

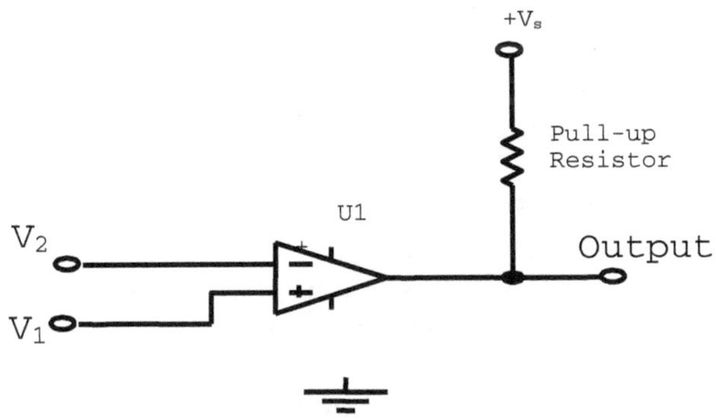

Fig. 2.14 Comparator with pull-up resistor

The advantage of the open-collector output configuration is that it is possible to use a different power source with different voltage levels for the output. It is also possible to connect the outputs of a several comparators in digital OR arrangement with only one pull-up resistor. This arrangement provides some flexibility.

In many process plants and other applications, the comparator is extensively used to generate alarm signals or taking some ON/OFF action. If the variable input is changing fast and it is free of high frequency noise, the above arrangement as is shown in figure 2.14 works perfectly alright. If the variable input has noise or it approaches the value of the reference voltage too slowly, the output of the comparator may switch back and forth between high and low values as the input is reaching the reference level. This quick transition of the output may create some detrimental effect on the switch, relay or on the equipment and it is better to be avoided. To avoid the problem, a hysteresis or dead-band window is provided at the reference level around which the output changes taken place as is shown in figure 2.15. A feedback resistor is connected between the output and the reference input (non-inverting input in the figure 2.15). The purpose is to change the value of the reference signal depending on the output condition. In absence of the resistance R_2, the value of the reference voltage is V. Usually the value of the feedback resistance R_2 is very large compared to the input resistance R_1.

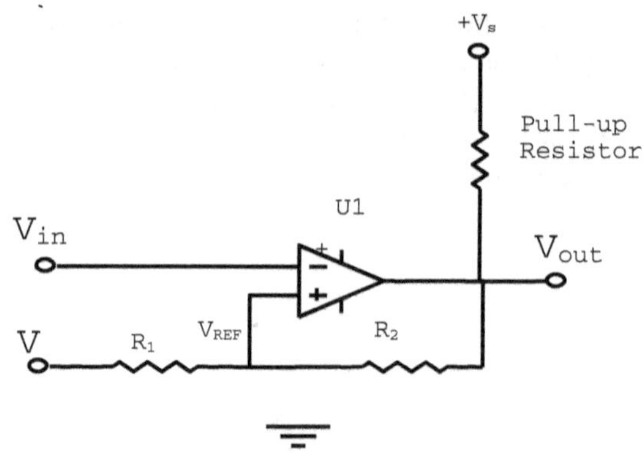

Fig. 2.15 Comparator with hysteresis

The expression of the modified reference voltage, V_{REF} with respect to the V_{out} and the earlier reference voltage, V, can be written as

$$V_{REF} = \frac{R_2}{R_1+R_2} V + \frac{R_1}{R_1+R_2} V_{out} \tag{2.16}$$

Since, $R_2 >> R_1$, the above expression can be approximated as

$$V_{REF} \cong \mathrm{V} + \frac{R_1}{R_2} V_{out} \tag{2.17}$$

When, V_{in} is less than V, V_{out} is HIgh (1), and the reference voltage is large and the higher value of the reference voltage is given by,

$$V_{REF} \cong \mathrm{V} + \frac{R_1}{R_2} V_{out} \tag{2.18}$$

Similarly, for V_{in} is greater than V, V_{out} is LOw (0), and the reference voltage is lower and the lower value of the reference voltage is given by,

$$V_{REF} \cong \mathrm{V} - \frac{R_1}{R_2} V_{out} \tag{2.19}$$

So the lower and higher V_{REF} gives a reference band. By choosing appropriate resistance R_1 and R_2, the hysteresis band can be properly designed. For proper calculation of resistor values, (2.16) is to be used.

Ex 2.8: A hysteresis comparator is to be designed for the following application. The system is used in a heating chamber with a negative temperature coefficient thermistor. The heating should be off when the temperature is reached 50°C. The system should not start again till the temperature falls to 45°C. The thermistor has a resistance of 10 kΩ at 50°C and 11 kΩ at 45°C. Assume the comparator works only with a 5V power supply.

Sol: There can be many ways of designing the hysteresis comparator. One solution is shown in figure 2.16.

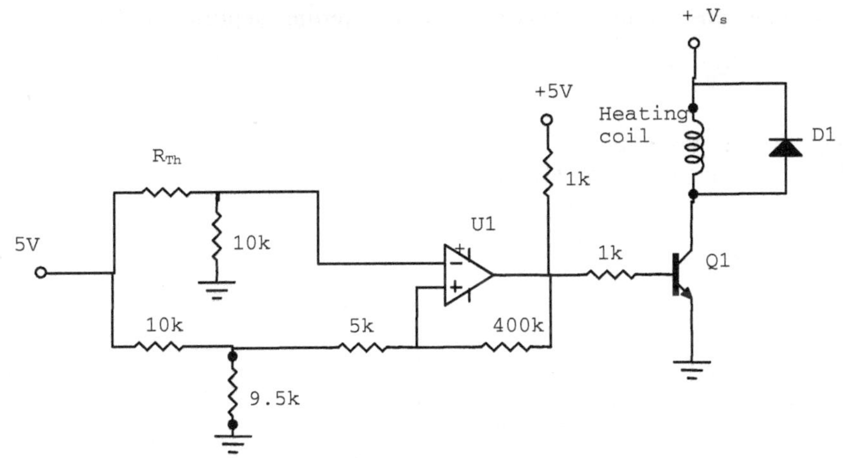

Fig. 2.16 Design of a hysteresis comparator for heating control application

2.10 A Few Guidelines to Design Signal Conditioning Circuit

Though it is not simple to have a set of guidelines applicable to each and every design but a few issues should be taken care for a successful signal conditioning system. If there is a choice of choosing the appropriate sensor, the sensor should be properly selected for the specified application. The sensor should be selected based on the criterion discussed in the earlier chapter.

Once the sensor is selected then the signal conditioning circuit is designed and implemented to interface the sensor to the processor. While the signal conditioning circuit is designed, the following guidelines are important:

Parameters: This is related to the type of output available from the sensor. Though the most common form of output is voltage but sometimes, current and frequency may also be available.

When the sensor is interfaced to a processor (microprocessor or microcontroller etc.), the Analog-to-Digital (ADC) converter is to be properly chosen. The hybrid controller may itself be properly chosen if the controller is used in a time-stringent application. The number of bits in the ADC or the resolution in the ADC conversion may be very important. Another important characteristic of the ADC is the conversion time of the ADC. Since the ADC takes some time to convert the analog input into digital form, the choice of microcontroller and the ADC converter should be based on the requirement. If there are more than one input to be converted with the help of multiplexor, this may be more severe requirement.

Ex 2.9: A displacement sensor is to be used to measure motion from 0 to 5 cm. The resistance changes linearly over this range from 0 to 10 kΩ. Develop a signal condition circuit to provide a linear 0 to 3.3 V output to directly interface to a microcontroller.

Sol: An inverting amplifier configuration is chosen as is shown in figure 2.17. By using the sensor at the feedback path of an inverting amplifier, we have

$$V_o = -\frac{R_s}{R_1} * (-5.1)$$

At $R_s = 10$ kΩ, Vo = 3.3 V.

So, $R_1 = (5.1 * 10)/3.3 = 15.15$ kΩ ≈ 15 kΩ.

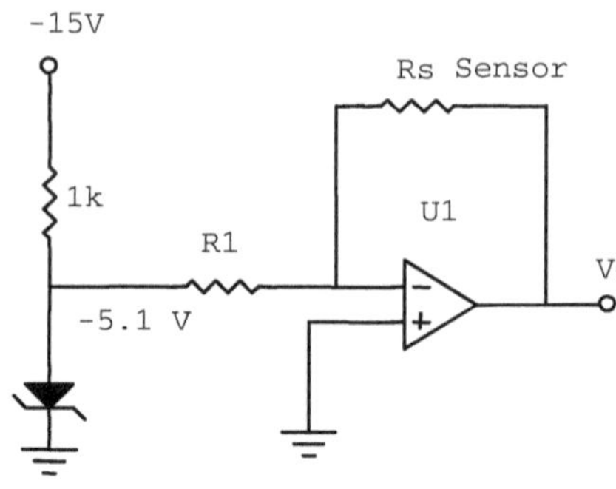

Fig. 2.17 Solution of the example 2.9

2.11 Factors Affecting Performance of Sensors

The performance of a sensor is influenced by many parameters. It is important to understand which parameters are affecting the performance and how it can be reduced or nullified. A knowledge of the characteristics of the sensors as a function of the influencing parameters is useful to design and develop necessary compensating electronic circuits. It is also important to note the degradation of the performance of the sensors with time. A common requirement is often heard from many when something goes wrong as "sensors should be calibrated" or "the sensor has been calibrated a long time back". It is important to understand the requirement of timely calibration of sensors as well as other components of the system to achieve the overall good performance of the sensing system.

2.11.1 Revisit of the Specification of Sensors

The user or designer selects a sensor based on the specified requirements. The sensor is chosen by consulting the specification of the sensor provided in the manufacturer's data sheet. In this section a re-visit of a few specifications will be done to emphasis the influence of a few parameters affecting the performance of the sensor.

2.11.1.1 Accuracy

For any sensor or any other element related to engineering design, it is expected that the error should be ideally zero or to a minimum value. The error is the difference between the actual value of the variable and the measured value. In practice, it is very difficult to achieve the ideal value. The term accuracy is specified to provide an idea how much the maximum error can be from a sensor. The accuracy is defined as, "The ability of a sensor to match the actual value of the quantity being measured". If in reality, the temperature is 25.0°C outside and a temperature sensor reads 25.0°C, then the sensor is accurate. The accuracy of a sensor is actually expressed as the inaccuracy and it can be expressed in different ways:

1. For many sensors, the accuracy is expressed as a percentage of a full-scale (FS). For example, for a sensor providing a 5 V FS reading with an accuracy of ±0.5%, there will be always an error or uncertainty of 0.025 V in the output.
2. For many sensors, the accuracy may be expressed as the percentage of actual reading. If the sensor reads a value of 1 V with an accuracy of ±1%, the sensor would have an inaccuracy of ±0.01 V.
3. Sometimes, the accuracy may be expressed as a percentage of the sensing range. Thus, for a temperature sensor measuring from 50°C to 100°C with an accuracy of ±2% in that range will have an inaccuracy of ±0.02 *(100-50)°C = ±1°C.
4. If the accuracy is expressed as an absolute value, for example, the accuracy of a temperature sensor is ±1°C, there would be an uncertainty of ±1°C in any value of temperature measured.

There is one more close term used along with the accuracy, which is precision. The precision is defined as, "The ability of a measurement to be consistently reproduced" and "The number of significant digits to which a value has been reliably measured". If on several tests the temperature sensor matches the actual temperature while the actual temperature is held constant, then the temperature sensor is precise. By the second definition, the number 2.1415 is more precise than the number 2.14.

Ex 2.10 A temperature sensor has a span of 50°C – 250°C. A measurement results in a value of 75°C for the temperature. Determine the possible readings and the errors in each case if the accuracy is (a) ±0.5% of FS, (b) ±1% of reading and (c) ±0.75% of the span.

Sol: From the discussion above, we have

(a) Error = ±0.005 * 250 = ±1.25°C. Thus the measured temperature may be anything in the range of 75 ±1.25 = 73.75°C to 76.25°C.
(b) Error = ±0.01 * 75 = ±0.75°C. Thus the measured temperature may be anything in the range of 75 ±0.75 = 74.25°C to 75.75°C.
(c) Error = ±0.0075 * (250 – 50) = ±1.5°C. . Thus the measured temperature may be anything in the range of 75 ±1.5 = 73.5°C to 76.5°C.

2.11.1.2 Overall Accuracy

In many situations, the sensor does not provide any electrical signal directly. It needs to be connected to an electronic circuit which is known as signal conditioning circuit to get the electrical output signal. The block diagram representation can be shown as in figure 2.18.

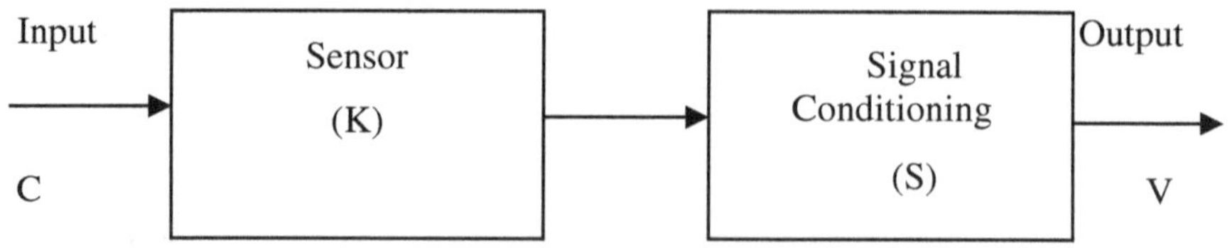

Fig. 2.18 Model of sensor for error estimation

The sensor can be represented as a transfer function between the output and input of the sensor. If the sensitivity or gain of the sensor is defined as K, we have

$$K = \frac{Output\ of\ the\ sensor}{input\ to\ the\ sensor} \tag{2.20}$$

For example, for a force sensor, the output is resistance in ohm (Ω) and the input is force in N, the sensitivity K will have a unit Ω/N. Similarly, the signal conditioning circuit may also be represented by another transfer function, S, as is shown in figure 2.18.

Now, each block will have an error or uncertainty. To analyse the effect of each and every component in a system, the overall error may be represented as is shown in figure 2.19.

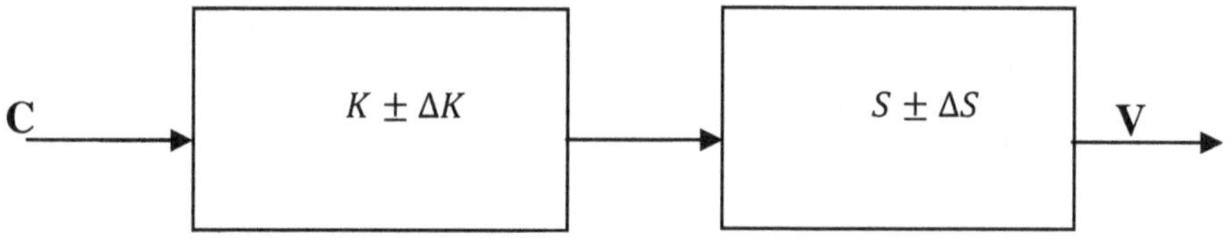

Fig. 2.19 Model of uncertainty in measurement using sensor

The output can be described as follows:

$$V \pm \Delta V = (K \pm \Delta K) * (S \pm \Delta S) \tag{2.21}$$

Where, V = Output,
ΔV = Uncertainty in output,
K and S = Nominal transfer functions,
ΔK and ΔS = Uncertainties in transfer functions,
C = Command or the variable.

Nominal output, $$V = K * S * C \tag{2.22}$$

From (2.21) and (2.22), the uncertainty in the output is obtained as,

$$\pm\Delta V = \pm\Delta K\, SC \pm \Delta S\, KC \pm \Delta K\, \Delta S\, C \tag{2.24}$$

So, we can write, $$\frac{\Delta V}{V} = \pm\frac{\Delta K}{K} \pm \frac{\Delta S}{S} \pm \left(\frac{\Delta K}{K}\right)\left(\frac{\Delta S}{S}\right) \tag{2.25}$$

It is assumed that the uncertainties with respect to the nominal value is very small,

$$\frac{\Delta K}{K} \ll 1 \quad \text{and} \quad \frac{\Delta S}{S} \ll 1.$$

So, the last term in (2.25) can be neglected and we have

$$\frac{\Delta V}{V} = \pm\frac{\Delta K}{K} \pm \frac{\Delta S}{S} \tag{2.26}$$

The (2.26) shows that the worst uncertainty in measurement is the sum of the individual uncertainties.

From statistical point of analysis, it may be more realistic to use the root-mean-square representation of system uncertainty. We can write,

$$\left|\frac{\Delta V}{V}\right|_{rms} = \pm\sqrt{\left(\frac{\Delta K}{K}\right)^2 + b\left(\frac{\Delta S}{KS}\right)^2} \tag{2.27}$$

Though the (2.27) provides an overall uncertainty slightly less than the worst-case value, it more likely reflects the actual value.

Sometimes, statistical analysis improves the measurement accuracy, especially when random error in measurement occurs. One method is to take many readings either at the same location or may be in a distributed places depending on the applications. Then the average of all the readings is taken as the actual value of measurement. This method improves the accuracy, though at the cost of time.

In real-time on-line measurement, this can be implemented as a moving window. For example, if the average is based on five readings, readings are taken as x(n), x(n-1), x(n-2), x(n-3) and x(n-4), x(n) being the latest reading. The average value is given by,

$$x_{av} = \frac{x(n)+x(n-1)+x(n-2)+x(n-3)+x(n-4)}{5} \tag{2.28}$$

When the new value of data is obtained, all old values are replaced by the new values by the following manner:

x(n-4) is replaced by x(n-3), x(n-3) is replaced by x(n-2) and so on with the latest value is x(n).

2.11.1.3 Standard Deviation

Other than average value, sometimes the standard deviation may also be useful to represent how the individual measurements are spread out about the average. The deviation of the individual measurement is calculated by:

$$d(1) = x(1) - x_{av}$$
$$d(2) = x(2) - x_{av}$$
..
,..
$$d(n) = x(n) - x_{av}$$

where d(1), d(2), .. d(n) are the deviations of individual measurement from the average value.

The standard deviation is calculated by

$$\sigma = \sqrt{\frac{d(1)^2+d(2)^2+d(3)^2 \ldots\ldots d(n)^2}{n-1}} \quad (2.29)$$

Ideally, the value of σ should be very small for good accuracy. The larger the standard deviation, the more spread out the measurements data.

Ex 10.2

Find the worst case uncertainty and statistically overall uncertainty for a sensing system with the uncertainty in the sensor is $\pm 1\%$ and for the signal conditioning is $\pm 0.5\%$.

Sol: From (2.26), we get $\frac{\Delta V}{V} = \pm\frac{\Delta K}{K} \pm \frac{\Delta S}{S} = \pm\left(\frac{\Delta K}{K} + \frac{\Delta S}{S}\right) = \pm (0.01 + 0.005) = \pm 0.015$

i.e. $\pm$ 1.5%.

Statistical approach, using (2.27), $\left|\frac{\Delta V}{V}\right|_{rms} = \sqrt{\left(\frac{\Delta K}{K}\right)^2 + b\left(\frac{\Delta S}{KS}\right)^2} = \pm \sqrt{(0.01)^2 + (0.005)^2} = \pm 0.0111$ i.e. 1.11%.

Ex 10.3 Temperatures in a room are measured in eight different locations as 10°, 25°, 21.2°, 22.1°, 19.7°, 18.5°, 27.1° and 20°C. Find the average temperature and the standard deviation.

Sol: $T(av) = \frac{19+25+21.2+22.1+19.7+18.5+27.1+20}{8} = 21.6°C$

The standard deviation is calculated as

$$\sigma = \sqrt{\frac{(19-21.6)^2 + (25-21.6)^2 + (21.2-21.6)^2 + \cdots ..(20-21.6)^2}{8-1}}$$

= 3.04°C.

2.12 Effect of Temperature

Temperature is an important factor to consider when dealing with sensors, as thermal condition may dictate the operating situation. This is especially more important experimenting with food materials, as the dielectric permittivity of the food products is a function of temperature. The interdigital capacitive type sensor was used to determine the fat and protein content in meat in a non-invasive way. The effective permittivity was calculated from the determination of the impedance of the sensor. The impedance of the sensor with meat sample as material under test (MUT) was measured for a range of temperatures. The experiment was run for nearly four hours, in which time the temperature of the meat sample changed from 2°C to around 12°C. It can be seen from figure 2.20 that the impedance decreases with increasing temperature. Around 15 minutes were taken for each sample during experimentation (experimentation done in a "temperature-controlled" room), and there was no appreciable change in temperature during this time. So, it can be concluded that if experiments were conducted in an environment in which the temperature is not maintained constant, the temperature will play a significant part in the results obtained.

In order to avoid the effect of temperature on the measurement, temperature needs to be measured and the temperature compensation should be included.

2.13 Degradation of Sensors

The performance of sensors usually becomes bad and then worse before it is completely out of action with time. Though the mechanism of the degradation is different for different sensors depending on the materials the sensor is made of, their construction, shapes, sizes, and applications. For sensors such as strain gauges used for measurement of strain or for the solid materials, it can be explained as follows:

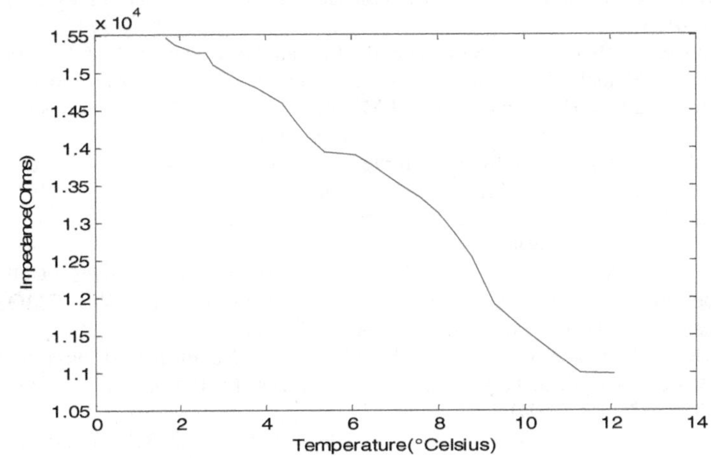

Fig. 2.20 Variation of impedance of the sensor with temperature

For solids consists of billions of atoms, when external force is applied, atomic spacing is adjusted to render the solid in equilibrium with all external forces acting on the object. The atomic spacing determines the physical dimension of the solid. With a change in the applied forces, the atoms in the solid re-arrange themselves again, to bring a new equilibrium situation with the new set of forces. With this re-arrangement, a change in the physical dimension takes place which is known as the deformation of the solid. While sensors are used to measure stress, strain or any other such parameters, the sensors are subjected to force on themselves. That forces may deform or change the physical structure of the sensors. Due to the deformation the sensors may not be able to respond in the normal way for which it was designed. This takes place with time of use and the performances of the sensors degrade with time. For some sensors, depending on the construction, material properties and applications, it may be possible to bring back the performance almost equal to the nominal state. But, for most of the situations, the changes taking place are permanent, the sensors may need to be replaced with new sensors.

2.14 Suggested Further Reading

The readers may read the following references for more knowledge and information.

1. Sen Gupta, G., Mukhopadhyay, S.C.: Embedded Microcontroller Interfacing: Designing Integrated Projects. LNEE, vol. 65. Springer (June 2010) ISSN 1876-1100, ISBN 978-3-642-13635-1
2. Saponara, S., Pasetti, G., Costantino, N., Tinfena, F., D'Abramo, P., Fanucci, L.: A Flexible LED Driver for Automotive Lighting Applications: IC Design and Experimental Characterization. IEEE Transactions on Power Electronics 27(3), 1071–1075 (2012)
3. Arcamone, J., Misischi, B., Serra-Graells, F., van den Boogaart, M.A.F., Brugger, J., Torres, F., Abadal, G., Barniol, N., Pérez-Murano, F.: A Compact and Low-Power CMOS Circuit for Fully Integrated NEMS Resonators. IEEE Transactions on Circuits and Systems—II: Express Briefs 54(5), 377–381 (2007)
4. Goes, J., Paulino, N., Pinto, H., Monteiro, R., Vaz, B., Garção, A.S.: Low-Power Low-Voltage CMOS A/D Sigma-Delta Modulator for Bio-Potential Signals Driven by a Single-Phase Scheme. IEEE Transactions on Circuits and Systems—I: Regular Papers 52(12), 2595–2604 (2005)
5. Hierlemann, A., Frey, U., Hafizovic, S., Heer, F.: Growing Cells Atop Microelectronic Chips: Interfacing Electrogenic Cells In Vitro With CMOS-Based Microelectrode Arrays. Proceedings of the IEEE 99(2), 252–284 (2011)
6. Wise, K.D., Anderson, D.J., Hetke, J.F., Kipke, D.R., Najafi, K.: Wireless implantable microsystems: High-density electronic interfaces to the nervous system. Proc. IEEE 92(1), 76–97 (2004)
7. Olsson III, R.H., Buhl, D.L., Sirota, A.M., Buzsaki, G., Wise, K.D.: Band-tunable and multiplexed integrated circuits for simultaneous recording and stimulation with microelectrode arrays. IEEE Trans. Biomed. Eng. 52(7), 1303–1311 (2005)
8. Kovacs, G.T.A.: Electronic sensors with living cellular components. Proc. IEEE 91(6), 915–929 (2003)
9. Cho, J.U., Le, Q.N., Jeon, J.W.: An FPGA-Based Multiple-Axis Motion Control Chip. IEEE Transactions on Industrial Electronics 56(3), 856–870 (2009)
10. Kimball, J.W., Flowers, T.L., Chapman, P.L.: Low-Input-Voltage, Low-Power Boost Converter Design Issues. IEEE Power Electronics Letters 2(3), 96–99 (2004)

11. Szlavik, R.B.: Strategies for Improving Neural Signal Detection Using a Neural–Electronic Interface. IEEE Transactions on Neural Systems and Rehabilitation Engineering 11(1), 1–8 (2003)
12. Kaigala, G.V., Behnam, M., Bliss, C., Khorasani, M., Ho, S., McMullin, J.N., Elliott, D.G., Backhouse, C.J.: Inexpensive, universal serial bus-powered and fully portable lab-on-a-chip-based capillary electrophoresis instrument. IET Nanobiotechnol. 3(1), 1–7 (2009)
13. Carnieletto, R., Brandão, D.I., Farret, F.A., Simões, M.G., Suryanarayanan, S.: Smart Grid Initiative. IEEE Industry Applications Magazine 17(5), 27–35 (2011)
14. DeHon, A.: Array-Based Architecture for FET-Based Nanoscale Electronics. IEEE Transactions on Nanotechnology 2(1), 23–32 (2003)
15. Pienimaa, S.K., Martin, N.I.: High-Density Packaging for Mobile Terminals. IEEE Transactions on Advanced Packaging 27(3), 467–475 (2004)
16. Willig, A., Matheus, K., Wolisz, A.: Wireless Technology in Industrial Networks. Proceedings of the IEEE 93(6), 1130–1151 (2005)
17. Roussy, A., Karthikeyan, S., Oomen, I., Ortlepp, T., Sujiono, E.H., Brinkman, A., Rogalla, H.: HTS Pulse-Stretcher and Second Order Modulator: Design and First Results. IEEE Transactions on Applied Superconductivity 15(2), 457–460 (2005)
18. Cárdenas, R., Peña, R., Pérez, M., Clare, J., Asher, G., Vargas, F.: Vector Control of Front-End Converters for Variable-Speed Wind–Diesel Systems. IEEE Transactions on Industrial Electronics 53(4), 1127–1136 (2006)
19. Mellodge, P., Vendetti, C.: Remotely monitoring a patient's mobility: A digital health Application. IEEE Potentials, 33–38 (March/April 2011)
20. Yin, M., Ghovanloo, M.: Using Pulse Width Modulation for Wireless Transmission of Neural Signals in Multichannel Neural Recording Systems. IEEE Transactions on Neural Systems and Rehabilitation Engineering 17(4), 354–363 (2009)

Chapter 3
Wireless Sensors and Sensors Network

Introduction

A sensor is a device that measures a physical quantity and usually converts it into an electrical signal which can be read by an observer or by an instrument. A sensor generates an electrical signal related to a physical, biological or chemical parameter.

A good sensor obeys the following rules:

(i) is sensitive to the measured property,
(ii) is insensitive to any other property,
(iii) does not influence the measured property.

The wireless sensors are an extension of traditional sensors with advanced learning and adaptation capabilities. The system must also be re-configurable and perform the necessary data interpretation, fusion of data from multiple sensors and the validation of local and remotely collected data. Wireless sensors therefore contain embedded processing functionality that provides the computational resources to perform complex sensing and actuating tasks along with high level applications.

The functions of a wireless sensor system can be described in terms of compensation, information processing, communications and integration. The combination of these respective elements allow towards the development of wireless sensors that can operate in a multi-modal fashion as well conducting active autonomous sensing. The wireless sensor is used as a sensor node in a wireless sensor network (WSN) system. In recent times there is a growing trend of using WSNs in different applications. A few applications are Precision agriculture, Environment Comfort & Efficiency, Smart Homes, Alarms, Security, Surveillance, Disaster Management, Health Care, Traffic Management, Transportation Safety, and Landmine Detection and so on. In fact, more and more new applications of WSN are reported in recent times.

The block diagram representation of a wireless sensor is shown in figure 3.1. The sensor is the most important element in this system as the main purpose is to measure the parameters of interest accurately and reliably over a long period of time. The sensor is interfaced to a processor as it requires the processing of the signal available from the sensors. The sensors may need some signal processing circuit before it is interfaced to the processor through the Analog-to-Digital

S.C. Mukhopadhyay: *Intelligent Sensing, Instrumentation and Measurements*, SSMI 5, pp. 55–69.
DOI: 10.1007/978-3-642-37027-4_3

converter and digital I/Os. The processor may need to store some processed or raw data obtained from the sensor. Finally in order to transmit the data some form of wireless communication circuit interface is required. The complete block diagram needs to be implemented with the help of electronic circuits to obtain a wireless sensor.

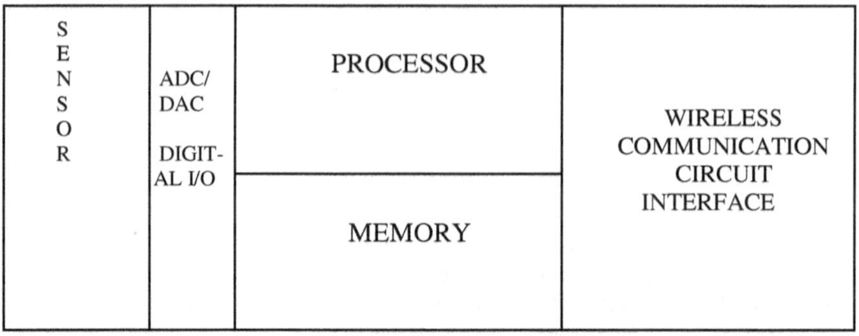

Fig. 3.1. Block diagram representation of a wireless sensor

The wireless sensors are the most important elements of a wireless sensor network (WSN). A schematic representation of a WSN is shown in figure 3.2. The wireless sensor form a node in the WSN and are distributed or deployed over a wide region to measure different parameters of interest. The measured data are transmitted either to the neighbouring node or to the coordinator and finally the data are collected by the central computer.

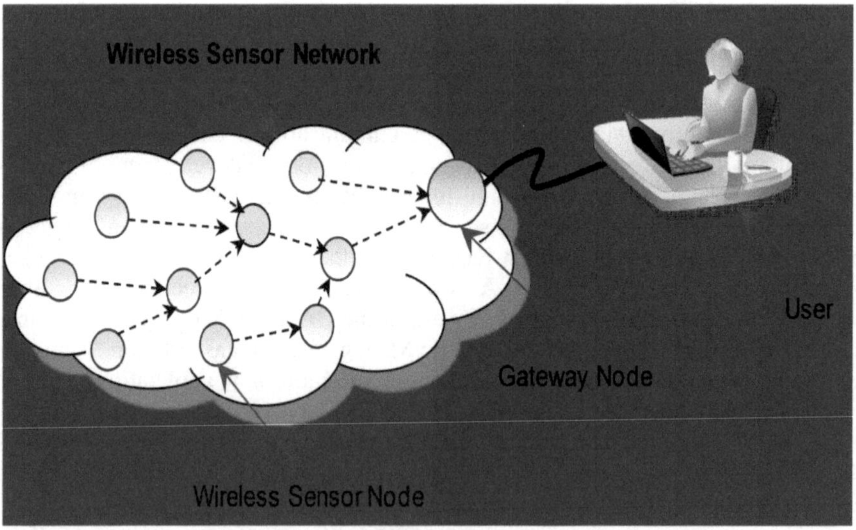

Fig. 3.2 Schematic representation of a Wireless Sensors Network (WSN)

3.1 Frequency of Wireless Communication

The wireless sensors communicate data over radio-frequency (RF) electromagnetic waves. It is very important for the WSN engineers to know the available radio frequency spectrum for the communication of sensors data. Usually a range of electromagnetic frequencies are reserved for the industrial, scientific and medical use and is known as ISM radio bands.

The ISM bands are reserved internationally for the use of radio frequency electromagnetic energy for industrial, scientific and medical purposes for communications and other applications. There are many applications in these frequency bands including radio-frequency process heating, microwave ovens, and medical diathermy machines. There can be electromagnetic emissions from these devices which can create electromagnetic interference and disrupt radio communication using the same frequency, so these devices were limited to certain bands of frequencies. The ISM bands are given in Table 3.1.

Table 3.1 Frequency of ISM band

Frequency range		Centre frequency	Availability
6.765 MHz	6.795 MHz	6.780 MHz	Subject to local acceptance
13.553 MHz	13.567 MHz	13.560 MHz	
26.957 MHz	27.283 MHz	27.120 MHz	
40.660 MHz	40.700 MHz	40.680 MHz	
433.050 MHz	434.790 MHz	433.920 MHz	Region 1 only and subject to local acceptance
863.000 MHz	870.000 MHz	866.500 MHz	Region 1 only and subject to local acceptance
902.000 MHz	928.000 MHz	915.000 MHz	Region 2 only
2.400 GHz	2.500 GHz	2.450 GHz	
5.725 GHz	5.875 GHz	5.800 GHz	
24.000 GHz	24.250 GHz	24.125 GHz	
61.000 GHz	61.500 GHz	61.250 GHz	Subject to local acceptance
122.000 GHz	123.000 GHz	122.500 GHz	Subject to local acceptance
244.000 GHz	246.000 GHz	245.000 GHz	Subject to local acceptance

The most commonly encountered ISM device is the home microwave oven operating at 2.45 GHz. However, in recent times, the same ISM band is very commonly used by the wireless sensors network for Bluetooth, zigbee, WiFi applications. The selection of communication frequency is very important and it should be such that is doesn't need any license to operate in any country.

Moreover, the interference with similar frequency of communicating signals should be as small as possible.

3.2 Development of Wireless Sensor Network Based Project

In this section the working experiences of the author will be described in order to provide a good picture to the readers about the necessary work behind the development of a successful WSN. The WSN can be simplified as the combination of two elements, one coordinator and the other is the sensor node, communicating via wireless medium. The sensor node measures the data accurately and reliably and transmits the data to the coordinator. The coordinator receives the data and processes it and does necessary post-processing such as storage, transmit, taking control action and so on. Usually the sensor node has its own identification (ID). In WSN, there are many sensor nodes, so the IDs of each sensor nodes will be different.

3.2.1 Wireless Sensor Based on Microcontroller and Communicating Device

The first system developed by the author and his group is based on microcontroller (Si-Lab C8051F020) along with the communicating device BiM-418-40. The BiM-418-40 is a Radiometrix make miniature low-power UHF module as is shown in figure 3.3 and is capable of half-duplex data transmission at speeds up to 40 Kbits/s over distances of 30 meters "in building" and 120 meters in open ground. The BiM-418-40 operates at an operating frequency of 418 MHz which is free to use in many countries but it may not be allowed in many countries too.

Fig. 3.3 Picture of the BiM-418-40

The block diagram representation of the coordinator module based on Microcontroller and Communicating Device configured on a personal computer is shown in figure 3.4. The microcontroller is connected to the computer and communicates through the RS-232 configuration. The connection between the microcontroller and the BiM-418-40 is done through the four pins, RX-Select (pin 16), TX-Select (pin 15), TXD (pin 14) and RXD (pin 12). It also needs an antenna to be connected at pin 2. Of course, the power supply and ground should be properly connected at the appropriate pins. It is to be noted that the chosen microcontroller, Si-Lab C8051F020 operates at 3.3 V and the BiM-418-40 needs 5 V power supply for its operation. In the block diagram of figure 3.4, the LED displays are optional.

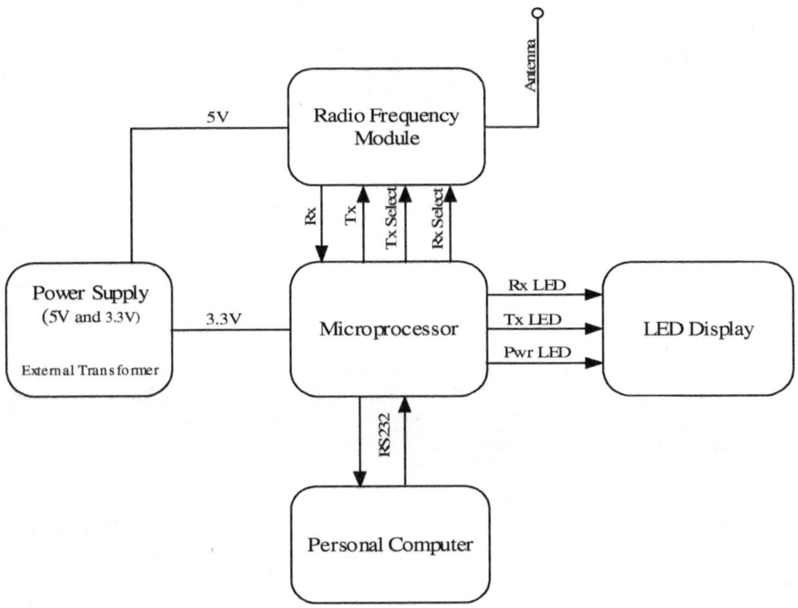

Fig. 3.4 Block diagram representation of the coordinator based on Microcontroller and Communicating Device configured on a personal computer

Based on the above description a coordinator system has been developed for the collection of the status of the different household appliances in a home monitoring system. The status (ON/OFF) of any household appliance is detected by a wireless sensor. The picture of the developed coordinator cum controller is shown in figure 3.5.

The communication between the PC or laptop and the microcontroller takes place through the COM port (RS232 port). A GUI (using any language Visual Basic, C#, or Java) is to be developed to read data from the RS232, using the instruction of "serialport", which has got several methods of handling the data. A few commands are:

Read Byte: Synchronously reads one byte from the SerialPort input buffer.

Read Char: Synchronously reads one character from the SerialPort input buffer.

Read Existing: Reads all immediately available bytes, based on the encoding, in both the stream and the input buffer of the SerialPort object.

Read Line: Reads up to the New Line value in the input buffer.

ReadTo: Reads a string up to the specified value in the input buffer.

Fig. 3.5 The developed coordinator cum controller for home monitoring system

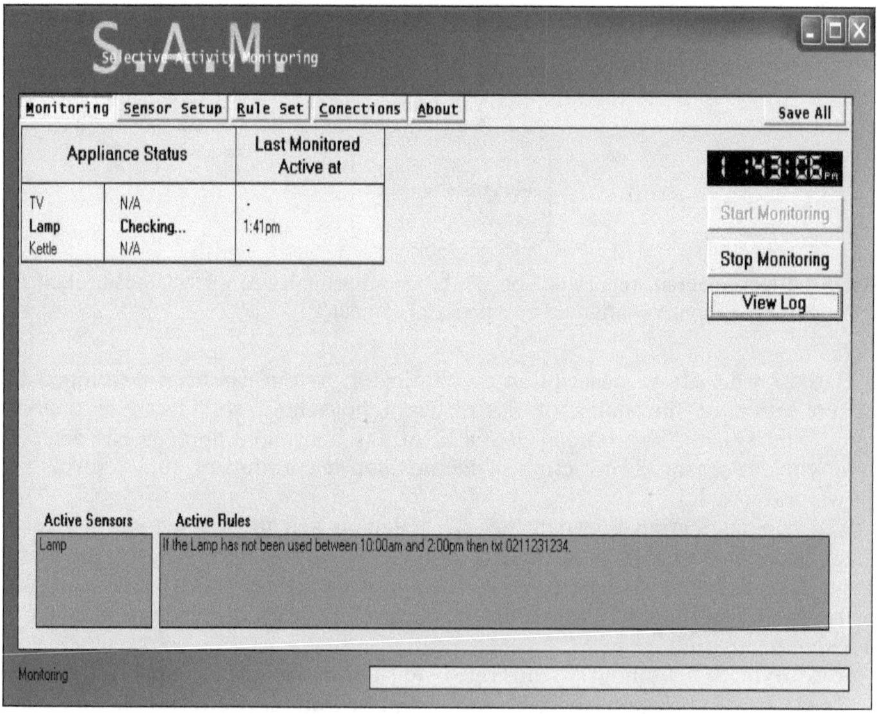

Fig. 3.6 The developed GUI for the home monitoring system

Write(String): Writes the specified string to the serial port.

Write(Byte(), Int32, Int32): Writes a specified number of bytes to the serial port using data from a buffer.

Write(Char(), Int32, Int32): Writes a specified number of characters to the serial port using data from a buffer.

Usually C# and Java are the mostly used programming languages to handle embedded systems, communications application programs. It is possible to store the data from the readline method to a variable inside the program and then it can be further stored in a file using file handling mechanisms of the respective programming language. A developed GUI for the home monitoring system to communicate with the coordinator is shown in figure 3.6. The detailed method of data handling has been discussed in chapter 5.

The development of a sensor node consists of similar components except the computer is replaced by the sensors. A sensor node for detecting an electrical appliance whether the appliance is ON or OFF is shown in figure 3.7. The sensor node is used in a home monitoring system to detect the status of an electrical appliances, bed, toilet, water-use and so on.

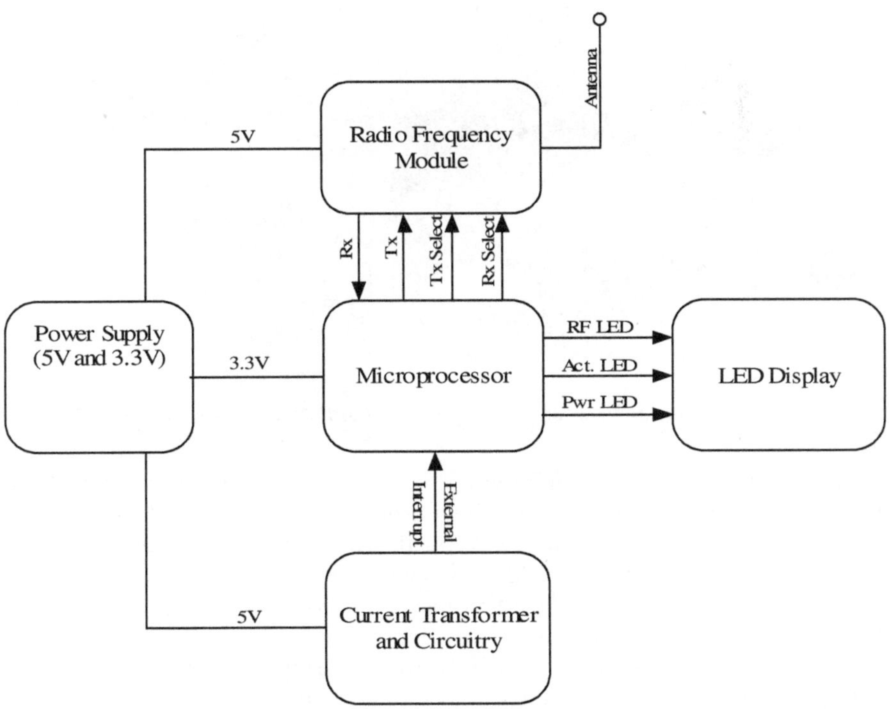

Fig. 3.7 The block diagram representation of a sensor node based on BiM-418-40

Based on the above block diagram a sensor node has been fabricated and is shown in figure 3.8. The system consists of microcontroller, communicating device BiM-418-40 and associated antenna, current transformer and its associated electronic circuit for detection of current flow.

Fig. 3.8 Fabricated sensor node for detecting the ON/OFF state of an electrical appliance

The communication protocol between the coordinator and the sensor node is based on an adhoc protocol. The controller program resides on the PC decides the communication. Whenever it wants to know the status of a sensor node, it initiates a command to the coordinator. The coordinator then transmits a request for the particular sensor node. The detailed steps are shown in figure 3.9. Due to the individual IDs of each sensor, only the desired sensor will respond. The communication data involved start byte, ID byte and stop byte while the request is made. When the data is obtained from a sensor node, it also involves the status of the device. It is to be noted that the sensor can only send data only when it is asked to do so by the controller.

There are some disadvantages of using the BiM-418-40 communicating device to develop a wireless sensor network. The disadvantages are:

1. It needs a microcontroller or a processor for its operation.
2. It is necessary to have a separate antenna for the communication of data. Consequently it needs more power and the cost of the system increases.
3. It requires a 5 V supply for its operation. This means a separate power supply is required as the microcontroller operates at 3.3 V.
4. The frequency of operation may not be licensed in many parts of the world. This means the system developed based on this frequency will not be allowed to run in many countries.
5. It needs more control for data communication. This will make the system slow. Moreover, the important data measured by the sensor may not be able to transmit to the coordinator as the communication takes place only when the coordinator asks the sensor node to do so.

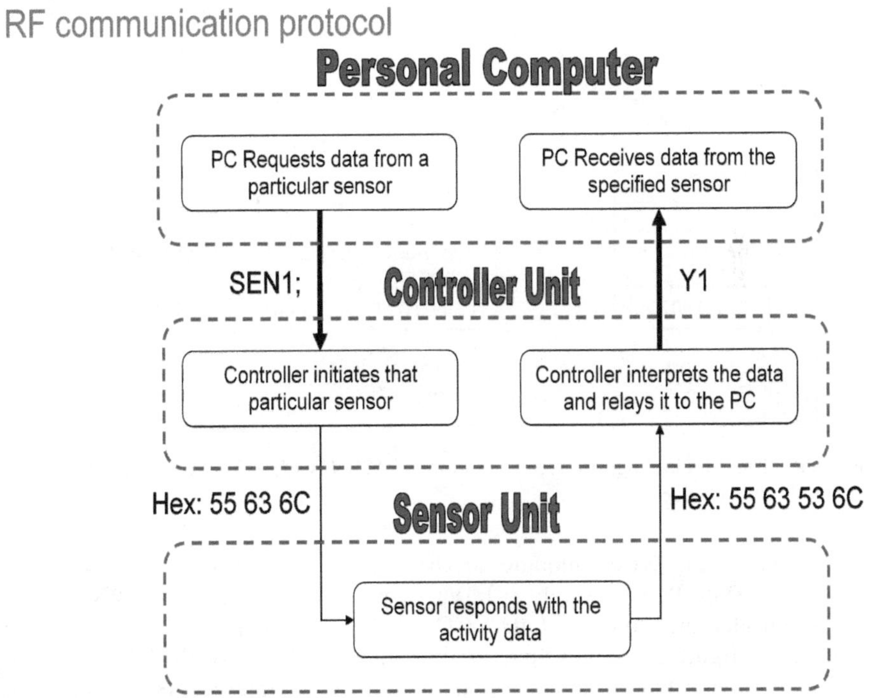

Fig. 3.9 The communication protocol for the data transfer between the coordinator and sensor nodes

3.2.2 Wireless Sensor Network Based on Microcontroller and Zigbee Communicating Device

Due to the above disadvantages of the BiM-418-40 communicating device, it is decided to develop the wireless sensor network based on a communicating device which operates at ISM band. In the next development stage Zigbee (also known as

Xbee) device operates at 2.4 GHz has been used to replace the BiM-418-40. The block diagram representation of the wireless sensors network based on microcontroller and zigbee communicating device is shown in figure 3.10. On the sensor node, the BiM-418-40 is replaced by the zigbee communicating device. On the receiver end/unit the Xbee co-ordinator receive data from various systems/ units operating at the time and stores the information on a database in the computer.

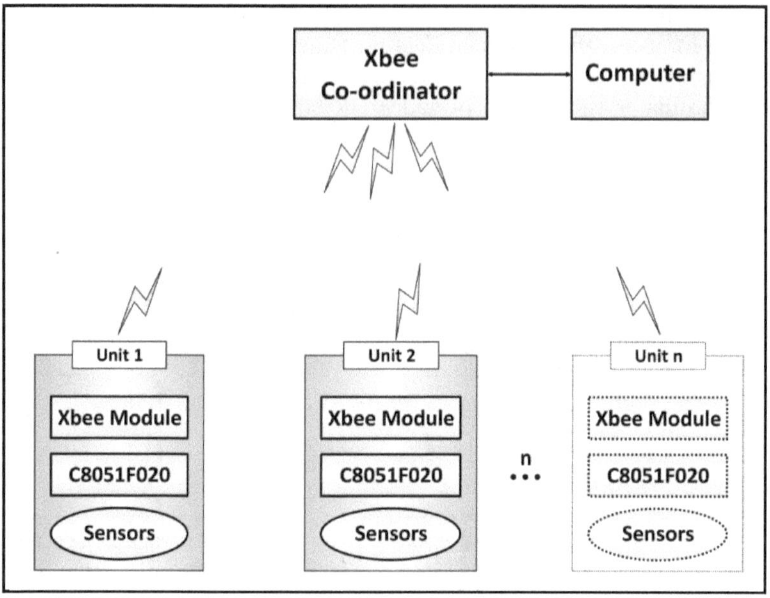

Fig. 3.10 The block diagram representation of WSN with microcontroller and zigbee communicating device

To connect the XBee module to the Microcontroller only four wires are necessary. The Power-Supply (3.3V), Ground and TX and RX of the Microcontroller are connected to VCC, GND, DIN and DOUT of the XBee module. The figure 3.11 shows pin configuration of XBee module. The Power and Ground are connected to the corresponding pins of the microcontroller. For serial communication Port 0 of the microcontroller is used. Pin 0 is TX and pin1 is RX for UART0. The UART0EN bit (XBR0.2) is set to logic 1, the TX0 and RX0 pins will be mapped to P0.0 and P0.1 respectively. Because UART0 has the highest priority, its pins will always be mapped to P0.0 and P0.1 when UART0EN is set to logic 1. If a digital peripheral's enable bits are not set to logic 1, then its ports are not accessible at the Port pins of the device. The crossbar assigns pins to all associated functions when a serial communication peripheral is selected (i.e. SMBus, SPI, UART). It would be impossible, for example, to assign TX0 to a Port pin without assigning RX0 as well.

Based on the above discussion a zigbee based sensing platform to measure the environmental parameters has been developed, the block diagram of the system is shown in figure 3.12.

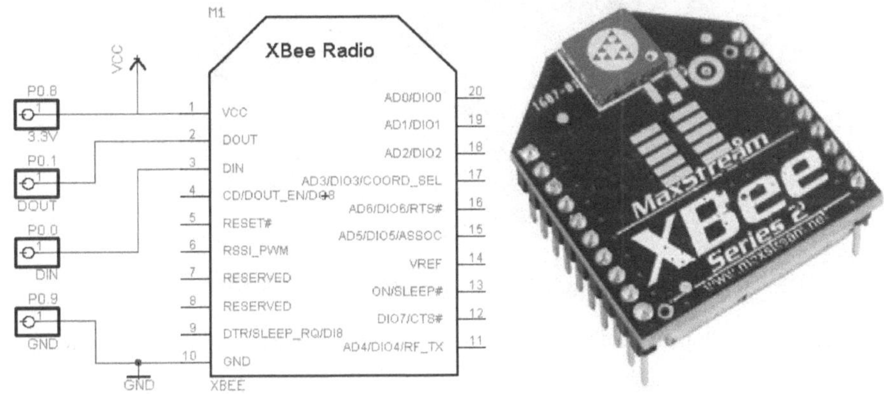

Fig. 3.11 The connection of the XBee with the microcontroller and its pictorial representation

A fabricated sensor node based on microcontroller and the zigbee is shown in figure 3.13. The system measures temperature, humidity, air pressure and light intensity.

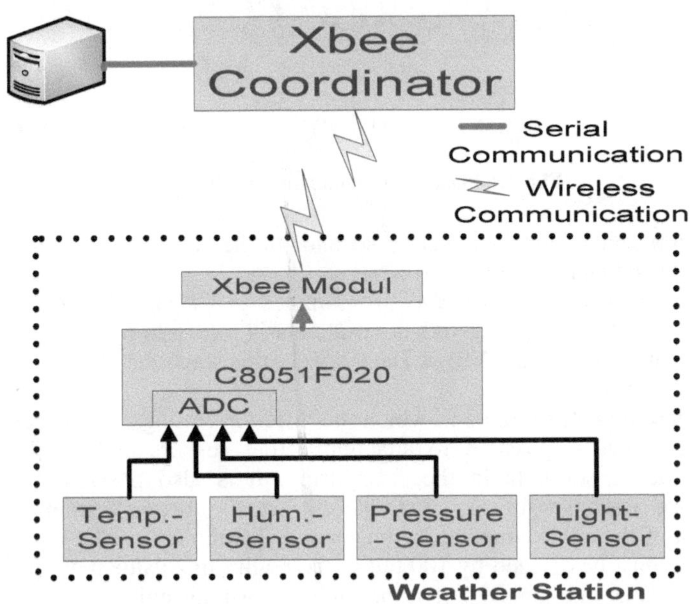

Fig. 3.12 Block diagram of the Environmental Monitoring System

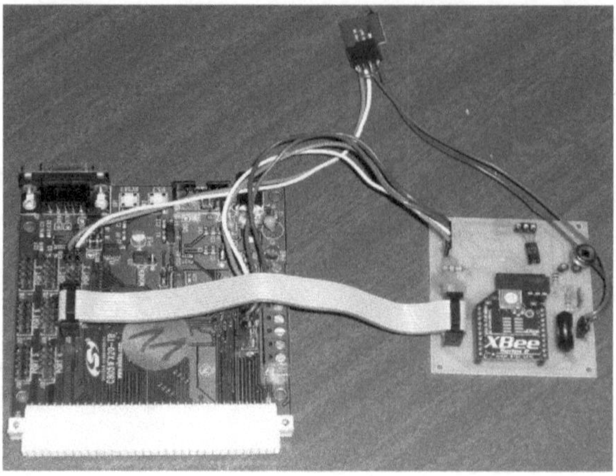

Fig. 3.13 Developed Weather Station connected to the C8051F020 Microcontroller

The transmission of the XBee Modules doesn't provide a checksum or any other possibility to verify the correctness of the received data. To avoid corrupted data and to see which station was sending the data an adhoc communication protocol is needed. The send string for the weather stations contained 27 characters as is shown in figure 3.14. The first three characters are the name of the station, then each sensor data are separated by a "-" character.

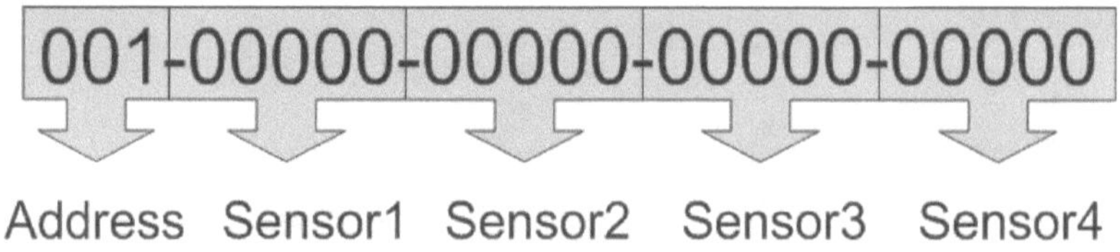

Fig. 3.14 Data Communication Protocol

Each station sends their data every 2 seconds to the coordinator where the data has to be collected and tested for correctness.

The Graphical User Interface was programmed in C# and captures the serial communication. The string received as serial data is split into 5 parts (the address and sensors) and saved in an Access Database. In this stage the GUI also tests the data for correctness.

The database contains a table in which the different settings for the sensors are stored. That makes it possible to attach different sensors and to change the position of the sensor data in the send string. It is also possible to make a calibration for each sensor, for example to get the better resolution in the measurement of the environmental parameters. To get the right value of the sensor the send data must be divided by 100 but in the reality this value may diversify to get the right value. In the GUI it is possible to display either one or multiple sensors as is shown in figure 3.15 to give the user the chance to show different graphs and to compare different stations.

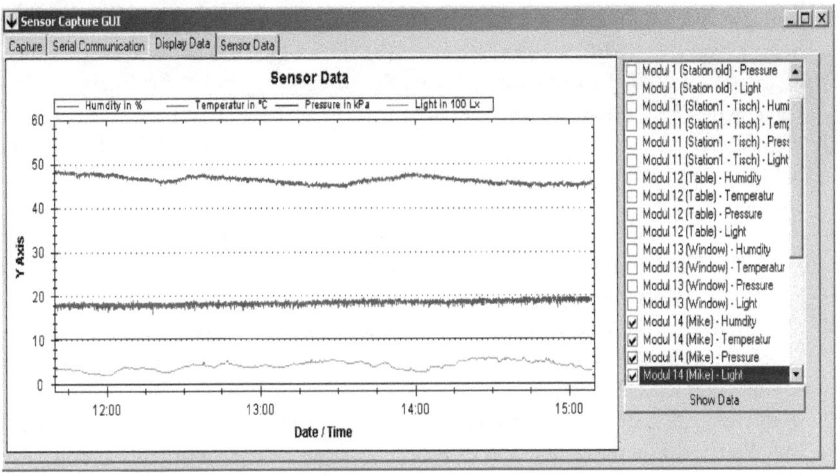

Fig. 3.15 GUI showing all 4 Sensors of one Weather Station

3.3 Wireless Sensor Network Based on Only Zigbee

It has been realized that there is no need to use a microcontroller in the sensor node if the node doesn't need any complex processing capability. The zigbee can directly collect the sensing data and configure the data to transmit it to the coordinator. But some steps are to be followed to configure the zigbee for this application.

The chapter 5 describes in details the necessary connection and programming to develop a zigbee based wireless sensor network system.

3.4 Suggested Further Reading

[1] Mukhopadhyay, S.C., Leung, H.: Advances in Wireless Sensors and Sensors Networks. LNEE. Springer (July 2010) ISSN 1876-1100, ISBN 978-3-642-12706-9

[2] Yick, J., Mukherjee, B., Ghosal, D.: Wireless sensor network survey. Comput. Networks 52(12), 2292–2330 (2008)

[3] Suryadevara, N.K., Mukhopadhyay, S.C.: Wireless Sensor Network Based Home Monitoring System for Wellness Determination of Elderly. IEEE Sensors Journal 12(6), 1965–1972 (2012)

[4] Suryadevara, N.K., Gaddam, A., Rayudu, R.K., Mukhopadhyay, S.C.: Wireless Sensors Network based safe Home to care Elderly People: Behaviour Detection. Sens. Actuators A: Phys. (2012), doi:10.1016/j.sna.2012.03.020

[5] Kaur, K., Mukhopadhyay, S.C., Schnepper, J., Haefke, M., Ewald, H.: A Zigbee Based Wearable Physiological Parameters Monitoring System. IEEE Sensors Journal 12(3), 423–430 (2012)

[6] Gaddam, A., Mukhopadhyay, S.C., Sen Gupta, G.: Elderly Care Based on Cognitive Sensor Network. IEEE Sensors Journal 11(3), 574–581 (2011)

[7] Lee, J.-W., Lee, J.-J.: Ant-Colony-Based Scheduling Algorithm for Energy-Efficient Coverage of WSN. IEEE Sensors Journal 12(10), 3036–3046 (2012)

[8] He, C., Kiziroglou, M.E., Yates, D.C., Yeatman, E.M.: A MEMS Self-Powered Sensor and RF Transmission Platform for WSN Nodes. IEEE Sensors Journal 11(12), 3437–3445 (2011)

[9] Alippi, C., Camplani, R., Galperti, C., Roveri, M.: A Robust, Adaptive, Solar-Powered WSN Framework for Aquatic Environmental Monitoring. IEEE Sensors Journal 11(1), 45–55 (2011)

[10] Lin, H.-C., Kan, Y.-C., Hong, Y.-M.: The Comprehensive Gateway Model for Diverse Environmental Monitoring Upon Wireless Sensor Network. IEEE Sensors Journal 11(5), 1293–1303 (2011)

[11] Mascareñas, D., Flynn, E., Farrar, C., Park, G., Todd, M.: A Mobile Host Approach for Wireless Powering and Interrogation of Structural Health Monitoring Sensor Networks. IEEE Sensors Journal 9(12), 1719–1726 (2009)

[12] Lee, J.-S., Cheng, W.-L.: Fuzzy-Logic-Based Clustering Approach for Wireless Sensor Networks Using Energy Predication. IEEE Sensors Journal 12(9), 2891–2897 (2012)

[13] Liu, X., Ma, Z., Liu, S., Cao, X.: Query Processing in Multi-User Scenario for Wireless Sensor Networks. IEEE Sensors Journal 11(10), 2533–2541 (2011)

[14] Jiang, H., Chen, L., Wu, J., Chen, S., Leung, H.: A Reliable and High-Bandwidth Multihop Wireless Sensor Network for Mine Tunnel Monitoring. IEEE Sensors Journal 9(11) (November 2009)

[15] Rodríguez-Sánchez, M.C., Borromeo, S., Hernández-Tamames, J.A.: Wireless Sensor Networks for Conservation and Monitoring Cultural Assets. IEEE Sensors Journal 11(6), 1382–1389 (2011)

[16] Song, L., Hatzinakos, D.: A cross-layer architecture of wireless sensor networks for target tracking. IEEE/ACM Trans. Network. 15, 145–158 (2007)

[17] Becker, T., Kluge, M., Schalk, J., Tiplady, K., Christophe: Autonomous Sensor Nodes for Aircraft Structural Health Monitoring. IEEE Sensors Journal 9(11), 1589–1595 (2009)

[18] Lhermet, H., Condemine, C., Plissonnier, M., Salot, R., Audebert, P., Rosset, M.: Efficient power management circuit: From thermal energy harvesting to above-IC microbattery energy storage. IEEE J. Solid-State Circuits 43(1), 246ff (2008)

[19] Tseng, H.-W., Sheu, S.-T., Shih, Y.-Y.: Rotational Listening Strategy for IEEE 802.15.4 Wireless Body Networks. IEEE Sensors Journal 11(9), 1841–1855 (2011)

[20] Niazi, M.A., Hussain, A.: A Novel Agent-Based Simulation Framework for Sensing in Complex Adaptive Environments. IEEE Sensors Journal 11(2), 404–412 (2011)

[21] Yoo, S.-E., Chong, P.K., Kim, D.-Y., Doh, Y.-M., Pham, M.-L., Choi, E.-C., Huh, J.-D.: Guaranteeing real-time services for industrial wireless sensor networks with IEEE 802.15.4. IEEE Trans. Ind. Electron. 57(11), 3868–3876 (2010)

[22] Ahn, H.-S., Ko, K.H.: Simple pedestrian localization algorithm based on distributed wireless sensor networks. IEEE Trans. Ind. Electron. 56(10), 4296–4302 (2009)

[23] Wang, L., Yang, G.-Z., Huang, J., Zhang, J., Yu, L., Nie, Z., Cumming, D.R.S.: A wireless biomedical signal interface system-on-chip for body sensor networks. IEEE Trans. Biomed. Circuits Syst. 4(2), 112–117 (2010)

[24] Han, D.-M., Lim, J.-H.: Design and implementation of smart home energy management systems based on zigbee. IEEE Trans. Consumer Electron. 56(3), 1417–1425 (2010)

[25] Pan, M.-S., Yeh, L.-W., Chen, Y.-A., Lin, Y.-H., Tseng, Y.-C.: A WSN-based intelligent light control system considering user activities and profiles. IEEE Sensors J. 8(10), 1710–1721 (2008)

[26] Turaga, P., Ivanov, Y.A.: Diamond sentry: Integrating sensors and cameras for real-time monitoring of indoor spaces. IEEE Sensors J. 11(3), 593–602 (2011)

Chapter 4
Power Supplies for Sensors

Introduction

The electronic components or devices in smart sensors and in sensor nodes of wireless sensor networks (WSN) require power for normal operation. In recent times wireless sensors and sensor networks have been widely used in many applications such as monitoring environmental parameters, monitoring and control of industrial situations, intelligent transportation, structural health monitoring, health care and so on. The advancement of electronics, embedded controller, smart wireless sensors, networking and communication have made it a possibility of the development of a low cost, low power smart wireless sensor nodes. Recently, interest in deploying WSN in different areas such as construction monitoring, security access control, lighting control, HVAC control has increased. Though a lot of researches are now conducted to minimize the power consumption of the sensor nodes, it is true with regard to sensor nodes that it requires energy to fulfill its desired role. Although there are certain situations where mains power might be available such as the smart homes, hydroponics green house, this will not always be the case.

The sensor needs power for operations of sensing, data processing and communication if it is used as a node in WSN. In WSN, most of the energy is required for data communication than normal sensing and data processing. For a simple comparison it can be said that the cost of transmitting 1 Kb of data over a distance of 100 metres is approximately the same as that used for the execution of 3 million instructions by a 100 million instructions per second processor. Usually wireless sensor nodes are typically very small electronic devices and power demand is also less, so they are equipped with a limited power source of less than 0.5-2 ampere-hour and 1.2-3.7 volts. Usually, the necessary energy is stored either in batteries or capacitors. For normal cases, batteries, both rechargeable and non-rechargeable, are the main source of power supply for sensor nodes. The batteries are also classified according to electrochemical material used for the electrodes such as NiCd (Nickel-cadmium), NiZn (Nickel-zinc), NiMH (Nickel-metal hydride), and lithium-ion. The batteries will get discharged with time due to normal operation of the sensor nodes, so energy harvesting techniques such as extracting energy from either solar sources, wind, temperature differences, or vibration may be exploited to renew their energy or recharge the battery. To extend the life of the batteries power saving policies may also be used. Dynamic

S.C. Mukhopadhyay: *Intelligent Sensing, Instrumentation and Measurements*, SSMI 5, pp. 71–90.
DOI: 10.1007/978-3-642-37027-4_4

Power Management (DPM) conserves power by shutting down parts of the sensor node which are not currently used or active. A Dynamic Voltage Scaling (DVS) scheme varies the power levels within the sensor node depending on the non-deterministic workload. It may be possible to obtain quadratic reduction in power consumption of the sensor node by varying the voltage along with the frequency.

There are many types of sensors available, they can be categorized into different groups depending on their operation in terms of power requirements. To extend the life of operation of the WSNs, it is recommended that it works with passive, omni-directional sensors. Each sensor node has a certain area of coverage for which it can reliably and accurately measure the particular quantity for which it has been deployed. The power consumption takes places in the sensor nodes are: signal sampling and conversion of physical signals to electrical ones, signal conditioning, and analog-to-digital conversion. Depending on the applications, the spatial density of sensor nodes in the field may be as high as 20 nodes per cubic meter.

4.1 Power Sources

In this section we will summarize different power sources available for WSN. Without power the sensor node or any other electronic based system will not operate. The powers available in different sources are described below:

4.1.1 Power from Mains Supply

In many applications the situations may allow to feed the sensor and electronics directly from mains supply. The switch mode regulated DC power adaptor includes multi-plug output adaptors and selectable polarity being small and lightweight may be a good choice for many applications. The general purpose power adapters will be available with the local electronic stores. One of such DC power adaptor is shown in Figure 4.1 which is available with the DickSmith Electronics (http://www.dicksmith.co.nz/product/M9926/international-power-adaptor-5-23v).

Fig. 4.1 Switch mode regulated multi-output DC power adaptor

The power adapter as shown in figure 4.1 can work with an input voltage variable from 90 V to 260 Volts AC and it has got a set of different DC output voltages, starting from 3 V to 12 V (3 V, 4.5 V, 6 V, 7.5 V, 9 V and 12 V). If the size of the power adapter looks big for the particular application, switched mode power converter may be designed to satisfy the requirements. Different small scale switching regulators are also available for consideration.

4.1.2 Battery

Batteries are commonly utilized as power sources for sensors especially for WSN nodes. However, batteries can emit a limited amount of energy before they are depleted. To address this issue, energy harvesting techniques may be employed. Energy harvesting allows nodes to replenish depleting energy sources of the batteries from external sources. Such energy harvesting helps to deal with the cost and problems associated with replacing and disposing of batteries. Various types of batteries are investigated in the following section.

4.1.2.1 Selection of Batteries

The selection of battery for a particular application depends on many factors. Though the availability and cost may be the most influencing factors but other factors such as size (volume), voltage and Ahr rating, life-time, environmental impact, memory effect and safety issues may also need to be considered. A few batteries of different sizes available off-the-shelf are shown in Figure 4.2.

Fig. 4.2 A few batteries available off-the-shelf

Usually two types of batteries are available to use: Disposable batteries (also known as primary batteries) are designed to be used once and discarded when depleted. The rechargeable batteries (also known as secondary batteries) are designed to be recharged and used multiple times. Usually the disposable batteries are cheaper than rechargeable batteries. In the following section, a few battery technologies are described.

4.1.2.2 Alkaline Batteries

One of the most commonly used household batteries is Alkaline Batteries accounting for 80% of the manufactured batteries in the US. It has an alkaline electrolyte of potassium hydroxide. The anode (negative terminal) is made of zinc powder, and the cathode (positive terminal) is composed of manganese dioxide for an alkanine battery. The nominal voltage of an alkanine battery is 1.5 V, a few of which can be connected in series for higher voltages. They are well known for their long shelf life. However, standard alkaline batteries are not suited for use in high-drain devices and are not rechargeable. There are alkaline rechargeable batteries, but they are generally of lower capacity and offer fewer recharge cycles than the other popular rechargeable batteries. One advantage of alkaline batteries is that in current design neither types contains toxic metals and both types can be disposed easily as these batteries are classified by the US Federal government as non-hazardous waste. In the state of California, USA, these batteries are to be disposed in accordance with the California Universal Waste Rules.

4.1.2.3 Lithium Battery

Lithium battery is another type of disposable battery. Lithium battery has got lithium metal or lithium compound as anode. There are different materials which are used as anode and similarly different possibilities for the electrolyte. A few materials used as cathode are Heat-treated manganese dioxide, Thionyl chloride, Sulfuryl chloride, Carbon monofluoride, Silver chromate, Copper(II) oxide, Copper sulfide, Iron sulfide and so on. Depending on the materials used as cathode different electrolytes are used, a few are: Lithium perchlorate in propylene carbonate and dimethoxyethane, Lithium tetrachloroaluminate in thionyl chloride, Lithium tetrafluoroborate in propylene carbonate, Lithium Perchlorate dissolved in Dioxolane and so on. Lithium is highly electropositive element. It is possible to develop very thin batteries using lithium. The specific energy in lithium battery can be very high compared to alkanine batteries. Though it is possible to get lower voltages than 3 V, the lithium batteries exhibit a stable voltage profile at 3 V and higher. Lithium button batteries as shown in Figure 4.3 are very popular in products such as portable consumer electronic devices. Lithium 9 V battery as shown in figure 4.3b is very popular for important devices like smoke detectors, wireless alarm systems, garage door openers etc. They are very reliable and provide longer life between battery changes and it boast an exceptional shelf life.

(a) (b)

Fig. 4.3 Lithium battery: (a) 3 V button type (b) 9 V type

4.1.2.4 Lithium-Ion Battery

Lithium-ion batteries are rechargeable batteries. During discharge the lithium ions (Li+) move from the negative electrode to the positive electrode. Usually, the positive electrode in Lithium-Ion battery is a metal oxide (may be either lithium cobalt oxide, a poly-anion such as lithium iron phosphate, or a spinel such as lithium manganese oxide). The negative electrode is made from carbon, (popular material is graphite) and the electrolyte is a lithium salt in an organic solvent. The Lithium-Ion batteries are very popular in consumer electronics and portable electronics and often used in high-drain devices such as laptops and cell phones. It has a good power to rate ratio. These batteries are usually recharged in specialized rechargers where safety requirements such as mandatory protection circuits are needed and internal protection circuits typically consume 3% of the stored energy. Figure 4.4 shows popular lithium ion batteries used in mobile phones and laptop computers.

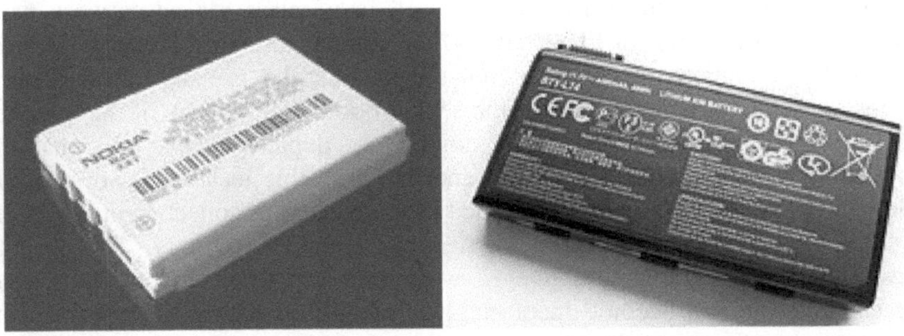

(a) (b)

Fig. 4.4 Popular Lithium-Ion battery for (a) Mobile phone (b) laptop

4.1.2.5 Nickel-Metal Hydride (NiMH) Battery

Due to harmful effect to environment Nickel Cadmium (NiCd) battery has been banned in many countries. Though the NiCd batteries are produced in many countries, the production will go down slowly. One of the alternatives to NiCd batteries are Nickel-Metal Hydride (NiMH) battery. NiMH battery use positive electrodes of nickel oxyhydroxide (NiOOH), like the NiCd, but the negative electrodes use a hydrogen-absorbing alloy. It is easy to dispose of as it does not contain toxic metals and isn't classed as a hazardous waste item. The disadvantages of NiMH are less voltage than alkaline batteries. The NiMH comes in different capacities where high capacity NiMH batteries may not charge completely in some chargers. Additionally they self-discharge at a high rate. The shelf-life is short and can be prone to steep drops in power when they are ready to be recharged.

4.1.2.6 Lead Acid Battery

Lead acid battery though may not be suitable for sensor applications in outdoor environment but it is one of the oldest rechargeable battery systems and is known for being quite economical in price, and rugged. Lead acid batteries are very popular in automotive applications due to their low cost and can supply high current required by automobile starter motors. Lead acid batteries should avoid deep discharges and should be charged more often. It can stay on charge with correct float charge. A disadvantage is that a deep-cycle battery delivers a 100–300 recharge cycles before it starts a gradual decline in capacity. Other disadvantages include size and weight of batteries and the toxic material contained.

4.2 Energy Harvesting

The sensors especially the sensor nodes used in wireless sensor network need continuous supply of power. The battery has limited amount of energy storage which will get depleted with use. This means that the primary or disposable batteries are not the viable option for the sensor nodes. The energy of the batteries should be replenished before it gets completely depleted. This tells that the rechargeable batteries along with some energy harvesting techniques are the only viable options. Energy harvesting is defined as the process of collecting energy from the surrounding environment and converting it into electricity. The energy harvesting techniques are gaining interest as a future next-generation energy source. In recent times, it has become increasingly attractive to extract energy from the ambient energy sources in the forms of light, wind, vibration, heat, radio waves, etc. Different methods to produce electricity from these different kinds of energy sources are now under development. Energy harvesting technology would help to eliminate the need for replacing disposable batteries and use of power cords. Different energy harvesting techniques are discussed in the following section and the pictorial representation is shown in figure 4.5. The advantages of energy harvesting can be summarized as:

i. Long lasting operability,
ii. No chemical disposal,
iii. Cost saving,
iv. Maintenance free,
v. No charging points,
vi. Inaccessible sites operability,
vii. Flexibility,
viii. Environmentally friendly,
ix. Applications otherwise impossible.

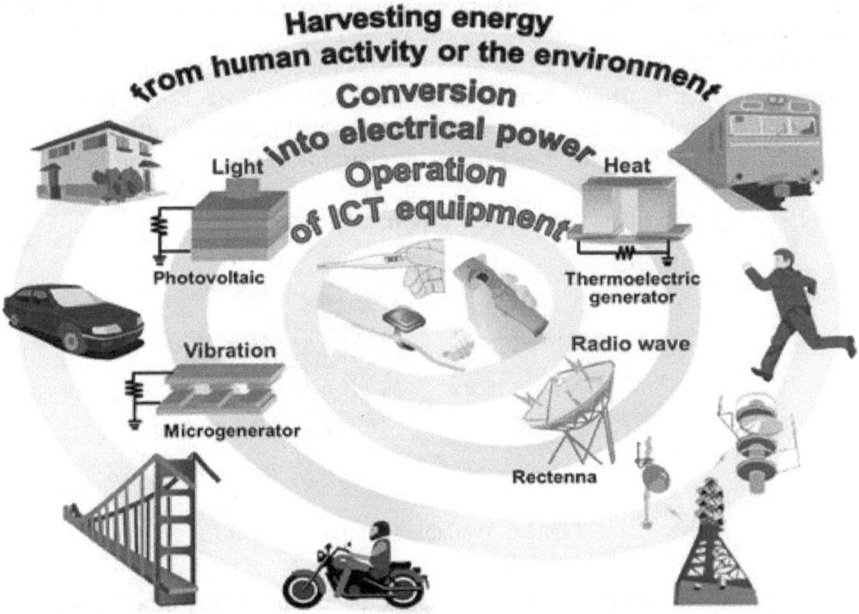

Fig. 4.5 Pictorial representation of energy harvesting techniques [Fujitsu Laboratories Ltd]

Though different techniques are explored separately but the amount of energy available from one type of sources may be limited. So it may be a good idea to combine a few sources together to form a hybrid harvesting of energy. A few challenges which are faced for this type of harvesting are:

i. Energy harvesting technologies are still not matured enough,
ii. The overall efficiency of harvesting system need to be improved,
iii. The harvesting system should meet the requirement of the size of the sensor node,
iv. The harvested energy need to be stored properly in an efficient manner.

4.2.1 Solar Energy

One of the most established energy harvesting methods is from solar energy, though still a lot of research on the development of high-performance solar panel is currently under investigation. Solar panels convert sunlight into electricity and have the ability to provide power to run most wireless sensor node applications. One of the common problems is that the power available from solar cells varies widely depending on the intensity of the sun. Another important consideration is that the sunlight is limited to a certain period of time on each day. So the storage of the energy in rechargeable batteries is the only available option. Additionally a number of other factors reduce the attainable power. Cloud cover and shadowing may block the sun's rays. So, a tracking system is required to follow the optimal angle of the sun.

A 12 Volt 1.26 Watt Solar Panel with the specification of Voltage (max): 18V-Current (max): Up to 70mA having Panel Size: 159 x 175 x 17mm is shown in figure 4.6a. The solar panel is ideal for charging sealed lead acid batteries.

A 12 Volt 4.5 Watt Solar Panel Ideal for charging sealed lead acid batteries is shown in figure 4.6b with the specification: Voltage (max): 18V, Current (max): Up to 250mA, Panel Size: 187 x 255 x 17mm.

(a) (b)

Fig. 4.6 Available solar panels

A simple experiment has been conducted to check the quality of available solar power panel and rechargeable batteries. A solar power battery charger (panel size is 75 mm X 55 mm) for rechargeable batteries to charge four batteries simultaneously has been used. The charger has been used to charge 1.2 V, 2500 mAhr Ni-MH rechargeable batteries. The purpose is to investigate how long the batteries can provide energy to a 2.5 W 90 Lumen LED power torch with the relation to the charging time. The components for the investigation are shown in figure 4.7.

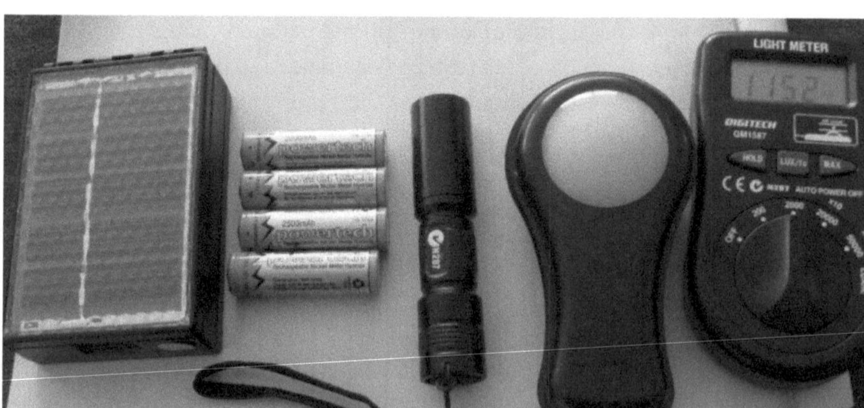

Fig. 4.7 A simple test for solar power battery charger

Figure 4.8 shows the bar graph of the discharge time of different Ni-MH rechargeable batteries as a function of charging time under the sunlight. The sunlight was not uniform and the charging took place at an ambient temperature varying between 14.5°C to 16.2°C and the relative humidity was between 58% to 62%. There was no appreciable wind during the experiment time. It is interesting to observe that the same type of batteries behave in a different way. Since it is impossible to maintain the sunlight uniform throughout the investigation the discharge time is not exhibiting a linear relationship with the charging time though

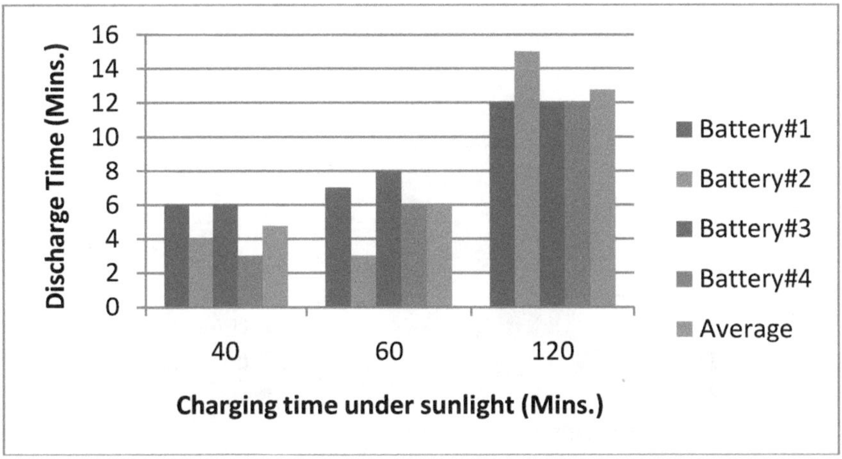

Fig. 4.8 Discharge time of Ni-MH rechargeable batteries as a function of charging time under sunlight

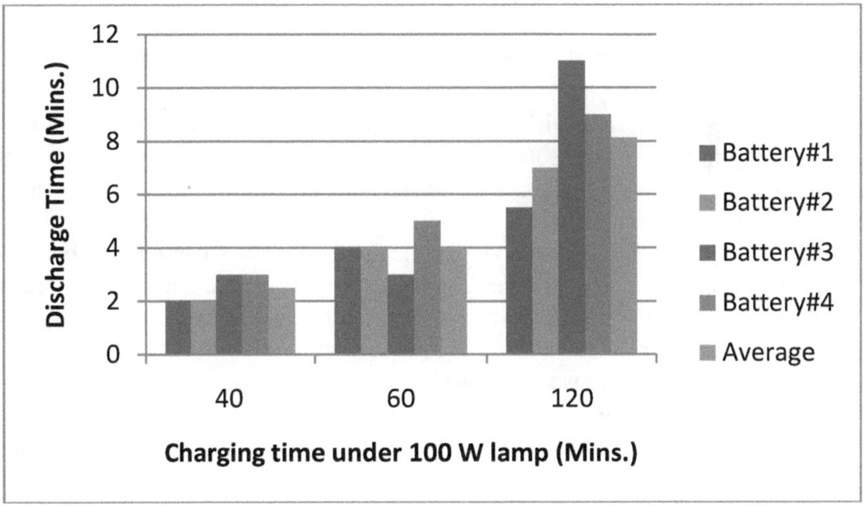

Fig. 4.9 Discharge time of Ni-MH rechargeable batteries as a function of charging time while the charger is placed under 100 W electric bulb

it is increasing with the charging time as it is expected. Figure 4.9 shows the bar graph of the discharge time of the same Ni-MH rechargeable batteries as a function of charging time under a controlled environment of 100 W lamp. It is seen that the sunlight produces better energy harvesting than the lamp which can be defined as ambient energy harvesting from home environment.

The charger has been placed under cloudy condition to investigate the energy harvesting ability. The figure 4.10 shows the comparison of three different conditions. It is seen that the cloudy condition is not very helpful from energy harvesting consideration.

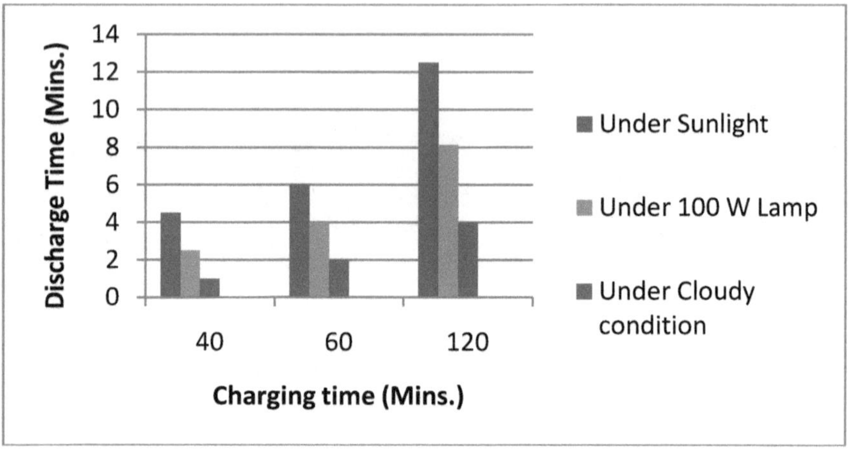

Fig. 4.10 Comparison of energy harvesting, under sunlight, under 100 W lamp and under cloudy condition

4.3 Further Investigation into Solar and Lead Acid Batteries

In this section some more results on the investigation of solar panel with lead-acid battery are presented. The solar energy is among the more feasible of the alternative energy harvesting options as promising results have already been achieved in the extraction of power from it. The solar cells are a technology which can play a key role in sensor node applications due to solar energy being a potentially limitless energy source. Lead acid batteries are one of the oldest rechargeable battery systems known for being quite economical in price, rugged, and forgiving if abused in particular a high over charge tolerance.

The solar panel test showed that during an average day, a voltage output of 15.44 V was attainable as is shown in figure 4.11 allowing for adequate energy supply to the lead acid battery which has a trickle charging voltage of around 13.6-13.8 V.

The specifications of the silicon solar PV cell are as follows:

Physical dimension: 254 mm X 294 mm,
Maximum Power Output: 5 W,
Maximum Generating Voltage: 16.8V,
Maximum Output Current: 0.30A.

Fig. 4.11 15.44V solar cell reading on an average day

In order to limit the voltage output of the solar panel to the lead acid battery, the TL431 was used where the output voltage can be set to any value between VREF 2.5 V and 36 V with two external resistors as is shown in figure 4.12 and the corresponding designed PCB is shown in figure 4.13.

The TL431 is stated as having active output, it provides a very sharp turn-on characteristic making it an excellent replacement for Zener diodes in many applications. This is of particular interest where the TL431 has benefits over Zener diodes for power regulation with regards to precision and voltage drop.

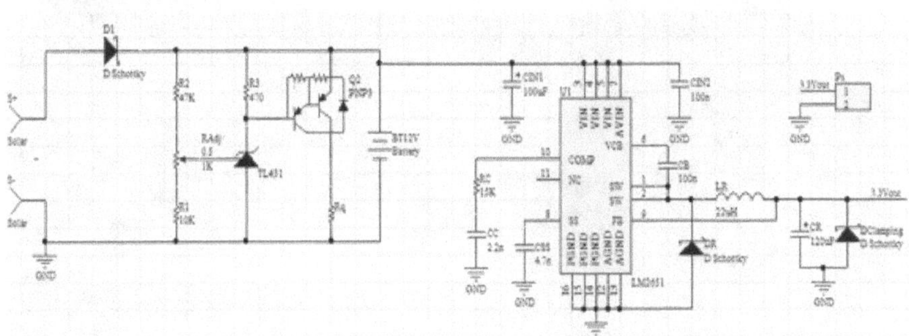

Fig. 4.12 Schematic Design of Solar Power Regulator to Battery and 3.3V Regulator

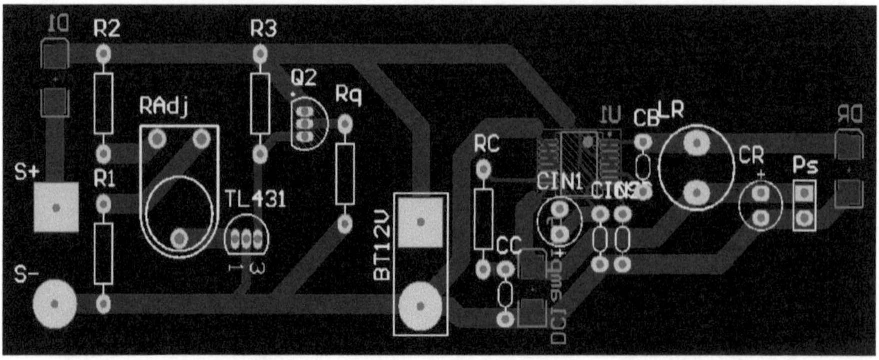

Fig. 4.13 PCB design of solar power regulator to battery and 3.3V regulator

$$V_{KA} = V_{REF} \times \left(1 + \frac{R_1}{R_2}\right) \quad (4.1) \qquad V_{KA} = V_{OUT}$$

Where $V_{Ref} = 2.5V$ and $V_{Out} = 13.7V$

Calculation for External Resistors

$R_1 = 4.48*R2$

Assuming $R_1 = 47$ KΩ gives $R_2 = 10.491$ KΩ.

Thus the selected values based on available resistor values are.

$R_2 = 10$ KΩ

$R_1 = 47$ KΩ.

However, the use of the standard resistor values chosen gives a voltage output of 4.202 V, as such an additional 1 KΩ pot is used to adjust/fine-tune voltage closer to the desired 13.7V.

Fig. 4.14 Testing Procedure with Solar Panel and Lamp as Light Source

Testing of the solar output voltage limiting system was executed by using the TL431 configuration. A solar panel with a lamp to provide the light source as is shown in figure 4.14 was used. The results of the Output Voltage limited to 13.7 V with TL431 are shown in figure 4.15.

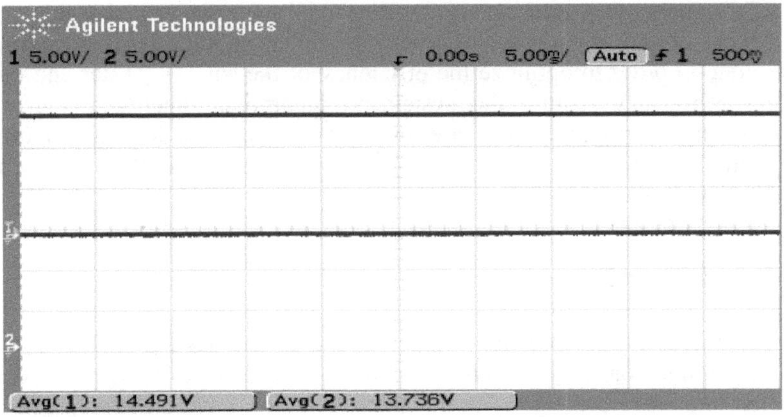

Fig. 4.15 Oscilloscope Reading for Circuit from Solar Panel where Output Voltage Limited to 13.7V with TL431

The investigation was performed in order to provide a 3.3V output voltage from the battery and solar system using a LM2651 1.5A High Efficiency synchronous switching regulator with the following features.

Key Features:

Ultra high efficiency up to 97% is shown in figure 4.16.

It operates at high efficiency over a 1.5 A to milli-amperes load range with 4 V to 14 V input voltage range.

Output voltage: 3.3 V.

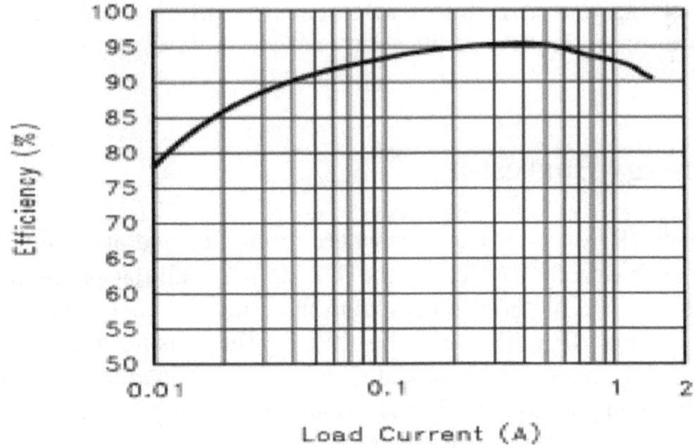

Fig. 4.16 Efficiency vs. Load Current LM2651

The LM2651 operates in a constant frequency (300 kHz), current-mode PWM for moderate to heavy loads, and it automatically switches to hysteretic mode for light loads.

The current work allows for charging of a lead acid battery with solar power should it perform adequately. The investigated system limits the voltage output from solar panel to lead acid battery in order to allow for its safe charging at 13.7 V. The work done is a basic solar harvesting implementation. Further exploration can be done in order to optimize the efficiency of the solar harvester and battery charging. This may require the system to incorporate intelligent electronics systems for monitoring, in which case further work should be done in weighing up the benefits of utilizing commercial offerings of plug-and-play solar energy harvesting modules.

4.4 Wind Energy

The wind energy is very popular in many countries and has established as one of the reliable non-conventional energy source. The kinetic energy of wind is converted into electric energy with the help of wind turbines consisting of blades, rotors, gears, generator and electronics. The availability of wind resources can vary based on geographical locations and topical features. The cost of wind energy is lower ($5-6/Watt) in comparison to some other alternative energy sources. The incremental improvements over the last few decades in wind technology have allowed for more reliable systems where wind turbines have generated as much electricity as conventional power plants worldwide. A disadvantage of wind power is the requirements for blades to be elevated in order to catch the force of the wind. Wind turbines have multiple moving parts where wear and tear can occur, also it can produce audible noise. Another disadvantage of wide energy is reliability factor as the strength of the wind is not constant and unpredictable. The wind energy is not suitable for using in sensors and sensor nodes due to huge amount of resource requirement.

The solar and wind are the two popular non-conventional energy sources currently supplying power. A lot of researches on different other techniques are currently under investigation to harvest energy. A few of them are discussed below.

4.5 RF Energy Harvesting

Radio frequency (RF) waves are everywhere around us and are emitted by energy sources such as television towers for the TV signals, cellular phone towers, radio towers and so on. The pictorial representation of the RF energy sources is shown in figure 4.17. This RF energy can be utilised for sensors and sensor nodes if sufficient amount of power is harvested. One of the applications based on this type of energy harvesting is passive RFID (Radio Frequency Identification Device) device. The RFID tag receives the wireless signal from the sensing device and utilised the power of that RF power to send back the stored information. For other cases where the sensing information is continuously required the challenge is to capture the free flowing energy using a power generating circuit and converting it into an usable direct current voltages. The converted DC voltages then can be stored either in a capacitor (or super-capacitor) or in a rechargeable battery for use or future use. The design challenge is to make the power generating circuit to a very small size.

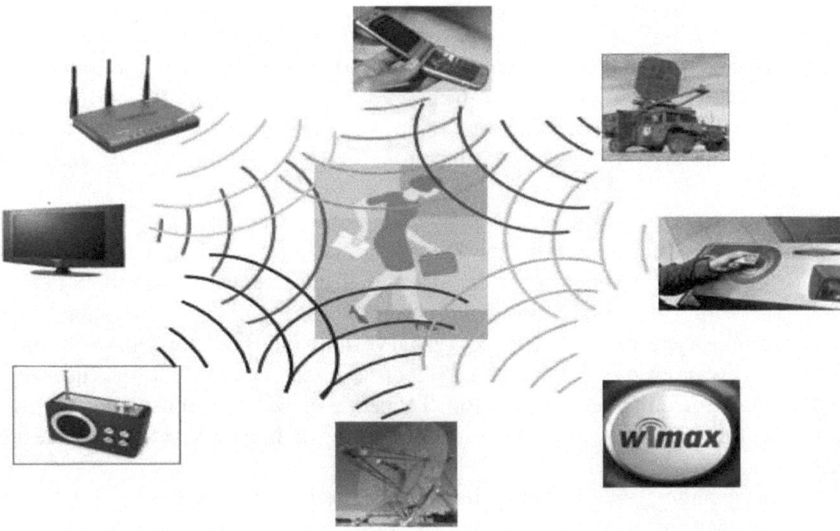

Fig. 4.17 Pictorial representation of the RF energy sources
(Source: D. BOUCHOUICHA, F. DUPONT, M. LATRACH, L.VENTURA, Ambient RF Energy Harvesting, International Conference on Renewable Energies and Power Quality (ICREPQ'10) Granada (Spain), 23th to 25th March, 2010)

A lot of researches are being currently reported on the harvesting of RF energy. The technological advancement due to which the amount of harvested power is slowly increased is through the development of smart receivers which can extract power from a broad range of frequencies. It is more challenging to extract energy if the receiver is located at a very far distance from the source of energy. The energy density is inversely proportional to the square of the distance from the source of the energy. So the receiver needs to operate in a very low power density region to produce a good amount of power.

The topic of RF energy harvesting is hugely important in the area of wireless sensors network. Since the solar power harvesting depends on the available of sunlight and the wind energy needs wind to flow, the placement of sensor node has an important role to play. The locations of the sensors may be such that sunlight may not enter and/or wind may not be allowed, the RF energy may be only viable option for that situation. The RF energy harvesting will be very useful if the sensors are located in remote, hazardous or sensitive area with continuous supply of energy. In the area of infrastructure monitoring or monitoring the health of structure the RF energy harvesting for supplying power to the wireless sensor nodes will play an important role.

4.6 Energy Harvesting from Vibration

The energy harvesting from vibration has been under investigation for quite some time. The mechanical energy involved with the vibration is converted into electrical energy and stored in storage device. The mechanical vibration is very common in day-to-day life though they vary widely in frequency and amplitude. Every system has its own natural frequency of oscillations. By utilizing the oscillations it is possible to harvest energy. Three different mechanisms are involved in this type of energy conversion: electromagnetic, electrostatic and piezoelectric.

In electromagnetic type of energy harvesting method the principle of electromagnetic induction is utilized. The magnetic fields need to be created either by using permanent magnet or electromagnet. The change of magnetic field is associated with the vibration involved in the harvesting system. The electromagnetic type mechanical energy harvester which converts the vibrations of steel highway bridges into stored electrical energy through the use of a translational electromagnetic generator. To power a wireless sensor network for monitoring the health of any structure is gaining a huge importance in recent times.

The electrostatic type of energy harvesting method is based on changing the capacitance resulting in from the vibration. For a simple expression the capacitor can be considered as a two-parallel plate capacitor in which the capacitor is inversely proportional to the separation distance. The energy stored is proportional to the capacitance value. The capacitor needs to be charged initially. Then with the help of mechanical vibration the capacitance can be changed and correspondingly the energy can be converted into electrical energy. The amount of energy harvested is related to the work done against the electrostatic force between the plates of the capacitor used. The advantage of working with electrostatic principle is that it is much easier to integrate with microelectronics and there is no need of having any smart materials. There is disadvantage too and that is the initial charge of the capacitor which means a voltage source is required for its operation.

The piezoelectric materials are famous for converting mechanical energy, usually from force, vibration or pressure into electrical energy. When a mechanical force is applied on a piezoelectric material, they are capable of producing electrical charge which is translated into a voltage difference between two surfaces. Due to this property piezoelectric material based energy harvester from vibration has gained a lot of popularity in recent times.

4.7 Thermal Energy Harvesting

The technology of thermoelectric generator or the generation of electricity based on temperature difference is quite old and is in use for many years. In 1821,

Thomas Johann Seebeck discovered that a thermal gradient formed between two dissimilar conductors produces a voltage and is known as Seebeck's effect. The technology was established and successfully used in spacecraft. It is also used in different unmanned surveillance sensors, tiny short-range communications devices etc. The thermoelectric generation has many good features characterized with vibration-less, light-weight, noiseless, maintenance free, small, and ability to work for long duration. If properly designed, the thermoelectric generators (TEGs) can be a good source of energy for sensor applications. In indoor applications, thermoelectric converters on the skin can provide more power per square centimeter than solar cells, particularly in adverse illumination conditions. Moreover, they work day and night [1]. The temperature difference between human skin and cloth has been utilised to generate sufficient energy which is used for a wrist watch. The disadvantage of thermal energy harvesting method is that for sufficient energy the temperature gradient should be sufficiently high and the low conversion efficiency.

4.8 Energy Management Techniques

Though different techniques are currently under investigation to harvest energy from different sources, still the amount of energy harvested is very small in amount, especially for sensor node in a wireless sensor network. In order to sustain the operation of the sensor nodes, different energy management techniques are also investigated in recent times. The purpose is to save energy for useful function and to reduce waste of energy for unimportant activities. A few techniques are discussed here [2].

4.8.1 Routing Protocol

A considerable research is currently under investigation to explore the most energy efficient routing protocol. The main objective is to save energy. Since larger the distance of the node from the coordinator more the amount of energy is required to transmit the data. In the energy efficient routing protocol the node need not to send the data to the coordinator rather it can send the data to the nearby node which is much closer to the coordinator. Some researchers have given name to one of the protocol as opportunistic routing protocol.

4.8.2 Introduction of Sleep Mode

To save energy some communication protocols allow introduction of sleep mode. The sleep mode is a special feature of the zigbee module. It is used to control the transmission rates of zigbee communication in order to maintain the battery power supply and hence to reduce the power consumption and thereby extending the battery life. This is achieved by defining the periodic sleep and wake cycles, for instance, during the sleep cycles the coordinator station holds the API data packets from the sleeping sensor station. When the sensor station wakes up from sleep, it transmits a poll request to the coordinator station and at this stage the coordinator can request data packets from the sensor station.

4.8.3 MAC Protocol

By efficient design of the medium access control (MAC) layer it may be possible to save significant amount of energy for normal operation and transmission. The MAC layer is responsible for the master node to poll every slave node for data transmission. A sensor node waste a significant amount of energy due to idle listening to the channel. Though it is similar to the sleep mode as discussed in the earlier section, MAC layer has many other abilities to configure the system.

4.9 Calculation for Battery Selection

It is important to choose a proper rechargeable battery for the given application. This section will provide some mathematical consideration while a battery is selected.

Let us take a practical situation with the following specifications:

Operating voltage of the sensors (V_S): 3.3 V

Current requirement of the sensor (I_n): 0.1 mA

Current requirement for transmitting data (I_{td}) : 20 mA

Time duration for transmission (t_{td}): 2 μs

Frequency of transmission (f_{td}): 1 kHz

Current requirement for transmitting data (I_{rd}): 18 mA

Time duration for transmission (t_{rd}): 1 μs

Frequency of transmission (f_{rd}): 2 kHz

Efficiency of the power converter to convert the battery voltage to the voltage suitable for sensor: 90%.

The battery voltage (V_B): 7.8 V.

Total power required for the sensor operation:

$P_S = V_S*I_n + V_S*I_{td}*t_{td}*f_{td} + V_S*I_{rd}*t_{rd}*f_{rd}$

$= (3.3 * 0.1*10^{-3} + 3.3 * 20*10^{-3} * 2*10^{-6} * 1*10^{3} + 3.3 * 18*10^{-3} * 1*10^{-6} * 2*10^{3})$ VA

= 0.5808 mVA.

The efficiency of the power converter is 90%, so the amount of required mVA from the battery is

0.5808 /0.9 mVA = 0.6453 mVA.

Given the battery voltage VB = 7.8 V, the discharge current required from the battery is 0.6453/7.8 = 0.0827 mA.

Assuming the supply of current from the battery is continuous (though actually it is not due to transmission and receive of data), and the battery voltage remains constant during the whole discharge process, the following table as shown in Table 4.1 can be prepared with respect to back-up time:

Table 4.1 Battery back-up time and required mAhr

Back-up Time (Hrs)	Required mAhr
1	0.0827
2	0.1654
4	0.3309
8	0.6618
16	1.323
24 (1 day)	1.9856
48 (2 days)	3.9712
72 (3 days)	5.9569
96 (4 days)	7.9425

4.10 Suggested Further Reading

The area of wireless sensor networks and energy harvesting is a vibrant research areas. The readers may read the following articles for more information:

[1] Leonov, V., Torfs, T., Fiorini, P., Van Hoof, C.: Thermoelectric Converters of Human Warmth for Self-Powered Wireless Sensor Nodes. IEEE Sensors Journal 7(5), 650–657 (2007)

[2] Niyato, D., Hossain, E., Rashid, M.M., Bhargava, V.K.: Wireless Sensor Networks with Energy Harvesting Technologies: A Game-Theoretic Approach to Optimal Energy Management. IEEE Wireless Communications, 90–96 (August 2007)

[3] Tutuncuoglu, K., Yener, A.: Optimum Transmission Policies for Battery Limited Energy Harvesting Nodes. IEEE Transactions on Wireless Communications 11(3), 1180–1189 (2012)

[4] Tashiro, K., Wakiwaka, H., Inoue, S.-I., Uchiyama, Y.: Energy Harvesting of Magnetic Power-Line Noise. IEEE Transactions on Magnetics 47(10), 4441–4444 (2011)

[5] Jornet, J.M., Akyildiz, I.F.: Joint Energy Harvesting and Communication Analysis for Perpetual Wireless Nanosensor Networks in the Terahertz Band. IEEE Transactions on Nanotechnology 11(3), 570–580 (2012)

[6] Jung, H.-J., Lee, S.-W., Jang, D.-D.: Feasibility Study on a New Energy Harvesting Electromagnetic Device Using Aerodynamic Instability. IEEE Transactions on Magnetics 45(10), 4376–4379 (2009)
[7] Nintanavongsa, P., Muncuk, U., Lewis, D.R., Chowdhury, K.R.: Design Optimization and Implementation for RF Energy Harvesting Circuits. IEEE Journal on Emerging and Selected Topics in Circuits and Systems 2(1), 24–33 (2012)
[8] Yoo, H., Shim, M., Kim, D.: Dynamic Duty-Cycle Scheduling Schemes for Energy-Harvesting Wireless Sensor Networks. IEEE Communications Letters 16(2), 202–204 (2012)
[9] Hande, A., Bridgelall, R., Zoghi, B.: Vibration Energy Harvesting for Disaster Asset Monitoring Using Active RFID Tags. Proceedings of the IEEE 98(9), 1620–1628 (2010)
[10] Beeby, S.P., Tudor, M.J., White, N.M.: Energy Harvesting Vibration Sources for Microsystems Applications. Meas. Sci. Technol. 17, R175–R195 (2006)
[11] Tan, Y.K., Panda, S.K.: Energy Harvesting From Hybrid Indoor Ambient Light and Thermal Energy Sources for Enhanced Performance of Wireless Sensor Nodes. IEEE Transactions on Industrial Electronics 58(9), 4424–4435 (2011)
[12] Gilbert, J.M., Balouchi, F.: Comparison of Energy Harvesting Systems for Wireless Sensor Networks. International Journal of Automation and Computing 05(4), 334–347 (2008)
[13] Marian, V., Allard, B., Vollaire, C., Verdier, J.: Strategy for Microwave Energy Harvesting From Ambient Field or a Feeding Source. IEEE Transactions on Power Electronics 27(11), 4481–4491 (2012)
[14] Liu, S., Lu, J., Wu, Q., Qiu, Q.: Harvesting-Aware Power Management for Real-Time Systems With Renewable Energy. IEEE Transactions on Very Large Scale Integration (VLSI) Systems 20(8), 1473–1485 (2012)
[15] Szarka, G.D., Stark, B.H., Burrow, S.G.: Review of Power Conditioning for Kinetic Energy Harvesting Systems. IEEE Transactions on Power Electronics 27(2), 803–815 (2012)
[16] Lopez-Lapena, O., Penella, M.T., Gasulla, M.: A Closed-Loop Maximum Power Point Tracker for Subwatt Photovoltaic Panels. IEEE Transactions on Industrial Electronics 59(3), 1588–1596 (2012)
[17] Manla, G., White, N.M., Tudor, M.J.: Numerical Model of a Non-Contact Piezoelectric Energy Harvester for Rotating Objects. IEEE Sensors Journal 12(6), 1785–1993 (2012)
[18] Wang, N., Zhu, Y., Wei, W., Chen, J., Liu, S., Li, P., Wen, Y.: One-to-Multipoint Laser Remote Power Supply System for Wireless Sensor Networks. IEEE Sensors Journal 12(2), 389–396 (2012)
[19] Knight, C., Davidson, J., Behrens, S.: Energy Options for Wireless Sensor Nodes. Sensors 8, 8037–8066 (2008)
[20] Giannakas, G., Plessas, F., Stamoulis, G.: Pseudo-FG technique for efficient energy Harvesting. Electronics Letters 48(9), 2 pages (2012)
[21] Richelli, A., Comensoli, S., Kovács-Vajna, Z.M.: A DC/DC Boosting Technique and Power Management for Ultralow-Voltage Energy Harvesting Applications. IEEE Transactions on Industrial Electronics 59(6), 2701–2708 (2012)

Chapter 5
Software Design for Data Reception and Analysis

Introduction

This chapter provides an overview of the complete software design necessary for data reception and analysis for a WSN based system. There are many ways to do wireless communication with sensors. ZigBee (or XBee) is extensively used and has been considered as the communication module. XBee is a wireless communication device from digi.com that uses ZigBee protocol. "ZigBee is a specification for a suite of high level communication protocols using small, low-power digital radios based on the IEEE 802.15.4-2003 standard for wireless personal area networks (WPANs), via short-range radio. The technology defined by the ZigBee specification is intended to be simpler and less expensive than other WPANs, such as Bluetooth. ZigBee is targeted at radio-frequency (RF) applications that require a low data rate, long battery life, and secure networking"[1].

It operates with 3.3V and uses 50mA (series 1). Some XBee pins can be set as Analog Input, Digital Input, Digital Output, or Analog Output (in terms of PWM) [2]. The numbers of pins are up to seven channels of analog input, nine channels of digital I/O and two channels of PWM. It is possible to send the measured sensor data to the central computer without the need of a microcontroller. The processing of data needs to be done at the central controller by this method.

The mechanism of sensor data reception is illustrated with an example in the following section:

A set of temperature, humidity and light intensity sensors are fabricated to be used as sensor nodes as shown in Figure 5.1. The sensors need some signal conditioning circuit before they are connected to zigbee. These are mesh-networked together to stream their data to a base station radio. This base station will be connected to a computer where the real-time measured data will be visualized on a computer screen.

The network topology of the sensing systems followed in this design is shown in Figure 5.2.

S.C. Mukhopadhyay: *Intelligent Sensing, Instrumentation and Measurements*, SSMI 5, pp. 91–117.
DOI: 10.1007/978-3-642-37027-4_5

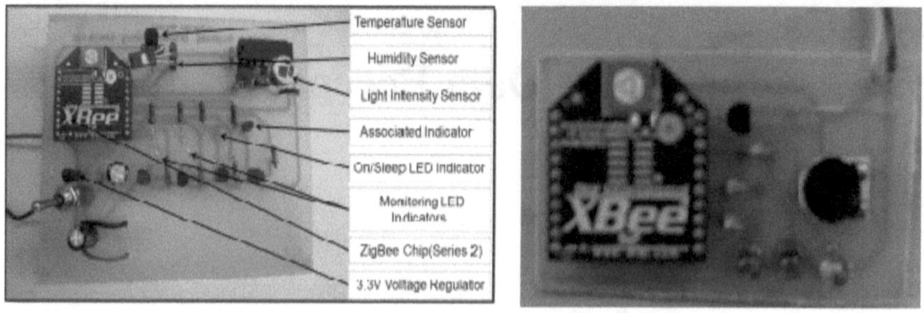

Fig. 5.1 Fabricated sensor node

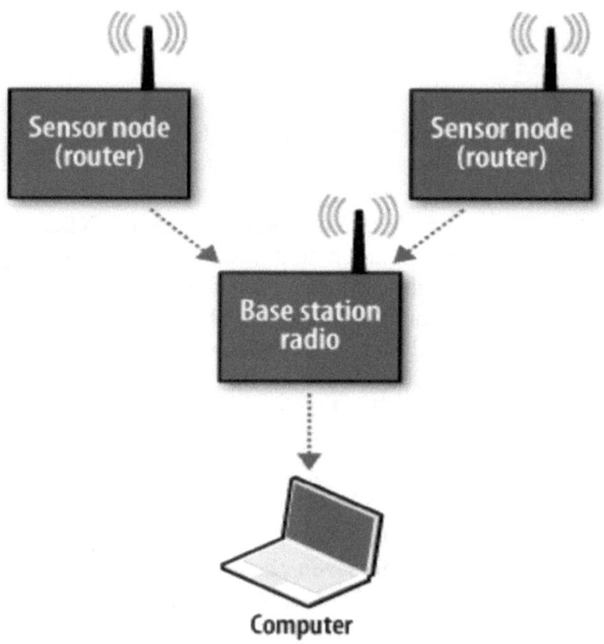

Fig. 5.2 Sensor Network topology followed in this project

5.1 Set-Up of the Wireless Sensor Network

The following section briefly describes the experimental setup performed in order to understand the mechanism of sensor data reception.

In this process, XBee's are configured one at a time. Firstly, a XBee module is labeled as coordinator to configure the one used locally (Coordinator) with computer. This is done by setting registers using "X-CTU (XBee *Configuration and Testing Utility*", a program provided by the manufacturer). It just listens and sends the data to the computer on a serial port. The remote configuration will tell the XBee that it should continuously transmit analog/digital sensor data to other

XBee's/coordinator in the area. A program is developed and installed on the computer that can read the serial data, store and further process the data according to the application.

Note: *In general, sensor network topology [3] can be any one of the following categories as shown in Fig.5.3:* ***Sensor data reception mechanism by the coordinator remains the same procedure.*** *To create a wireless network of ZigBee, at-least two nodes should be involved including coordinator node and sensor node type (Router/End device) to be able to communicate and work in PAN (Personal Area Network).*

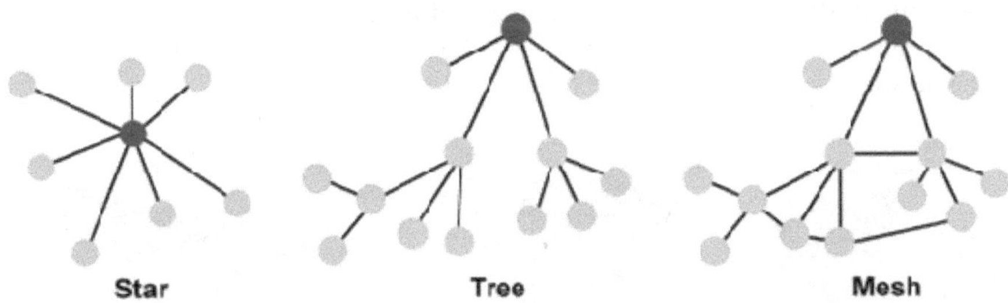

Fig. 5.3 Sensor network structures. ● Coordinator (Base Station), ○ Sensor device end or Router.

The setup of the project for sensor data reception is shown in figure 5.4. This setup is used for illustration and understanding the mechanisms such as receiving data, storage and communication with ZigBee sensor nodes.

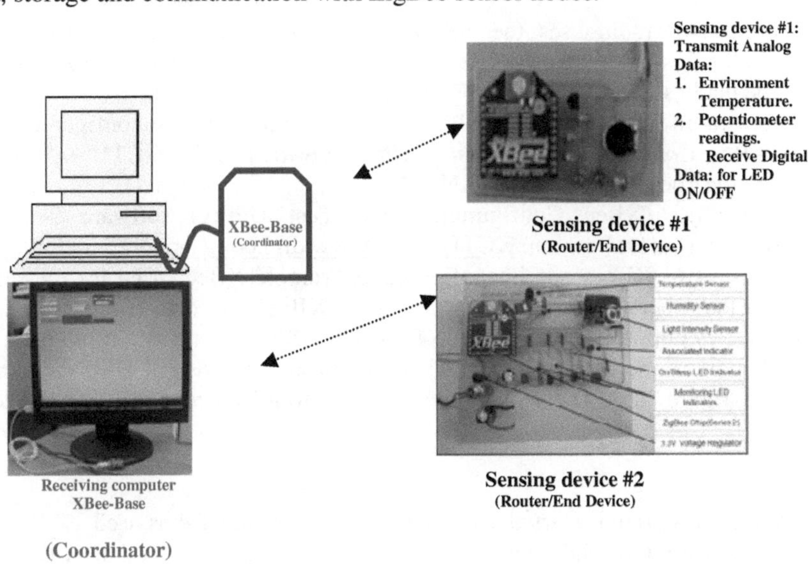

Fig. 5.4 Sensor data reception setup

5.2 Steps to Configure the ZigBee Radio Modules

The following steps are followed to program the complete system.

5.2.1 ZigBee Explorer USB

The most important element for setting up the ZigBee communication protocol is to have the ZigBee Explorer USB. The pictorial representation of the ZigBee Explorer USB is shown in figure 5.5. Both the coordinator Zigbee module and the sensor node ZigBee module are configured using the ZigBee Explorer USB. The coordinator ZigBee module remains connected to the computer through the USB port for data reception while in operation. The steps which are followed to setting up the network are discussed in the following section.

Fig. 5.5 ZigBee explorer USB for setting up ZigBee network

5.2.2 Preparation of the Coordinator (Base Station Radio)

The coordinator radio uses the API (*Application Programming Interface*) [4] firmware for this project to work because I/O data is only delivered in API mode. It is to be made sure to select the API version for the coordinator.

Once a radio has been set to API mode it can then only be configured in X-CTU (*XBee Configuration & Test Utility software*). The X-CTU is used to configure the coordinator with a PAN_ID.

The X-CTU (XBee Configuration and Test Utility) software is to be downloaded: (Latest Version: XCTU http://www.digi.com/support)

To set up two XBees one is configured as "remote XBee" used for the sensor node and the other is configured as the "local XBee" used as coordinator. The local XBee is popped into the "XBee Explorer USB" and then it is plugged into the USB port of the computer. This is done through the use of USB cable as is shown in figure 5.5. The following steps are followed to configure the local ZigBee:

Steps:

1. X-CTU Program is started and Modem configuration Tab is used.
2. The Modem type and Functional Set of the Local ZigBee module with values are set:

Modem = XBEE XB24-B,
Functional Set: ZNet 2.5 Coordinator API, Version: 1147.

3. The register ID-PAN ID with 64-bit address is set (Ex: 1234).
4. Each network is defined with a unique PAN identifier (PAN ID).
5. The register BD-Baud Rate is set with 9600.

[Any value can be selected; but the values should be the same for both the coordinator and sensor-node].

6. The register IR-IO sampling rate is set with 14 (in Decimal 20 ms).

[This can be configured for different sampling rates].

7. The "Always Update Firmware" is checked. "Write" tab to be pressed for the new configuration to store it on the XBee and waited for the write process to complete successfully.
8. Read option is clicked to check the parameter values are stored in the registers correctly. The Coordinator SL:________________ and SH:____________ values for reference are noted.
9. The XCTU Program is closed.

The screenshot for the local ZigBee configuration is shown in figure 5.6. The coordinator ZigBee is popped out from the explorer and the second ZigBee to configure sensing device Radio module is popped in.

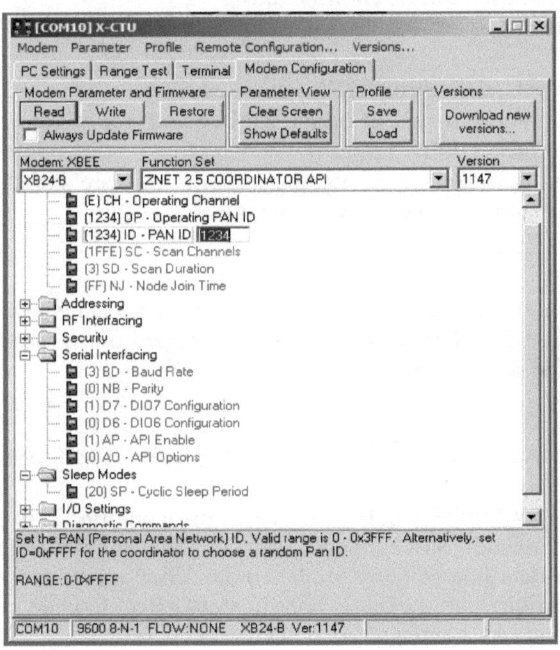

Fig. 5.6 The screen-shot of the coordinator configuration

5.2.3 Configuring the Remote ZigBee (Sensing Device #1 Radio Module)

The remote ZigBee (sensor-node module) is popped in into the USB Explorer. The following steps are followed to configure the module:

1. X-CTU Program is started and modem configuration Tab is used.
2. The Modem type and Functional Set of the Local XBee module are set with values: Modem = XBEE XB24-B,
   Functional Set: ZNet 2.5 Router/End Device AT
   Version: 1247
3. The register ID- (PAN ID) is set with 64-bit address (Ex: 1234).
4. The registers DL and DH of remote ZigBee are set with the Serial LOW (SL) and HIGH (SH) of the Coordinator.
5. The registers D0, D1 to 2 that tells ZigBee to read these pins as analog inputs are set. The number of inputs to be used depends on the specific application.
6. The registers D2 to 4 that tells ZigBee to read these pins as Digital inputs are set. This depends on the specific application.
7. The register BD-Baud Rate is set with 9600. This must be the same value what has been used for the coordinator.
8. The register IR is set for the sampling rate of every 20ms (IR = hex 14 = decimal 20).
9. The "Always Update Firmware" is checked. "Write" tab to be pressed for the new configuration to store it on the ZigBee and waited for the write process to complete successfully.
10. Read option is clicked to see the parameter values are stored in the registers correctly and Note: Coordinator SL:________________ and SH:____________ values for reference.

The XCTU Program is closed. The screenshot for the local ZigBee configuration is shown in figure 5.7. The remote ZigBee is popped out from the explorer and popped in into the sensing unit as shown in Fig. 5.4.

5.3 Brief Description of API Mode Data Transmission

The XBee 802.15.4 Modules have two modes of operation, Transparent and API modes [5, 6]. By default, XBee modules start in transparent mode; the coordinator sends necessary data immediately to all remote XBee modules (sensor-nodes) to establish connection, but the base coordinator XBee module responds to the terminal command (AT commands).

In API mode, all serial data is ignored unless it is in API command. Using API commands it is possible to send AT commands to the base and remote XBee module or send and receive serial data from a remote XBee module. The figure 5.8 shows the frame structure of API command and the figure 5.9 shows the frame structure of the response.

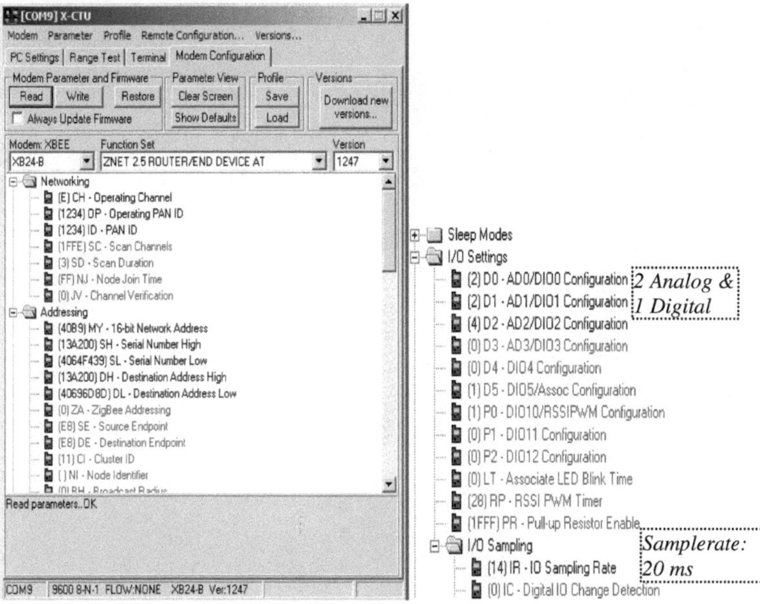

Fig. 5.7 The screen shot for the remote ZigBee configuration

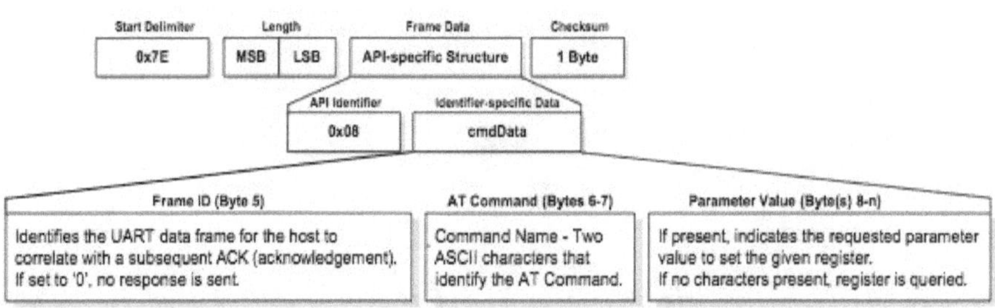

Fig. 5.8 AT command API frame structure [2]

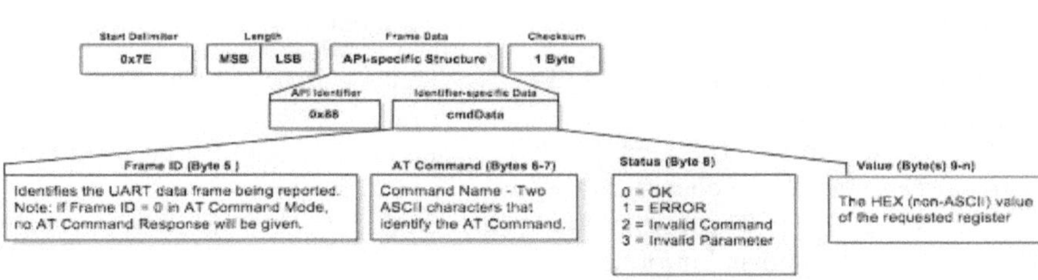

Fig. 5.9 AT command response frame structure [2]

The figure 5.10 shows the packets received by the coordinator.

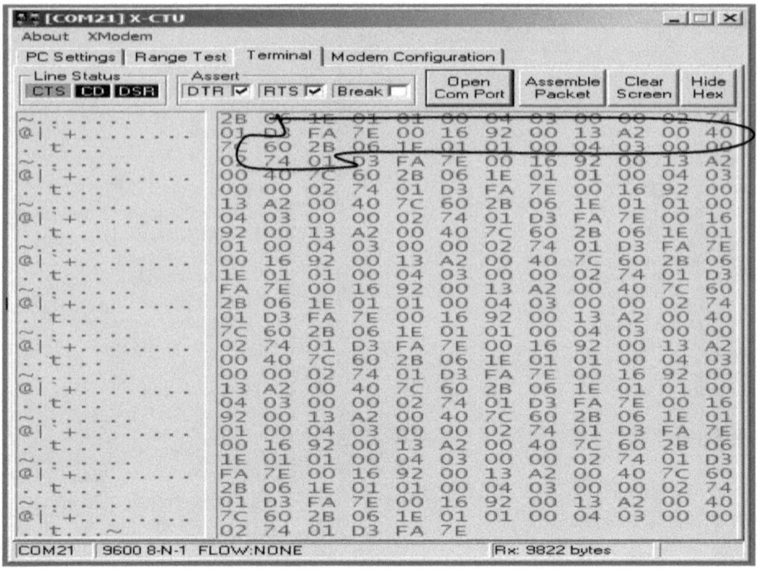

Fig. 5.10 X-CTU terminal displaying sensor Hex values

The following are the contents of a packet received at the coordinator.

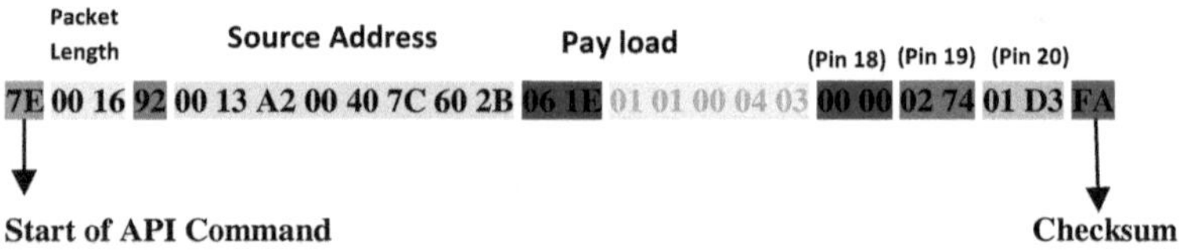

Note: XBee frames contain Hex values. Therefore, XBee packets contain Hex values. Number of Bytes received by the coordinator depends on the number of channels (pins) configured at the sensor device.

A brief description of the packet structure is given below:

Start of API command: Every API packet begins with 0x7E. Anything before 0x7E is ignored.
Command Length: The next two bytes contain the length of the rest of the packet, excluding the checksum.
Command Body: The rest of the bytes, excluding the checksum, contain specific API command information.
Checksum [5]: The sum of all bytes after the command length, including the checksum should be 0xFF.To calculate the checksum, all the bytes after the command length except the least significant byte are summed up. The sum is complemented to get the checksum.

Remote AT command request: All remote AT command requests begin with 0x17.
Frame ID: If the frame ID in an API command is non-zero, then the response will contain the ID.
64-bit address: This is the 64-bit XBee module address. For mesh-network configuration, it becomes compulsory to have the 64 address for exchanging data.
16-bit address: This is the 16-bit XBee module address. For mesh networking this address is not required. If 64-bit addressing is being used, and the 16 bit address is not used then this field should be filled with FF FE.
Apply changes flag: If this byte is 0x02, then changes take effect immediately on the remote XBee module.
AT command: This is the AT command used to the remote XBee module.

Note: In some cases a sensor-node may send data repeatedly, for example, the collected sensor values. It may not be important that some data is missed by the coordinator (controller) as long as a recent sample is obtained. In other cases it may be more critical that each transmission is accepted and used by the controller. Sometimes data may be a single byte. Other times it may be a collection of values that need to be accepted in the proper order.

In order to receive continuous sensor data as shown in figure 5.10, an effective, robust, flexible and easy to use program has been designed and developed as shown in figure 5.11.

The following section explores basics of designing and developing a GUI for sensor data reception, storage and transmitting digital value to sensor device:

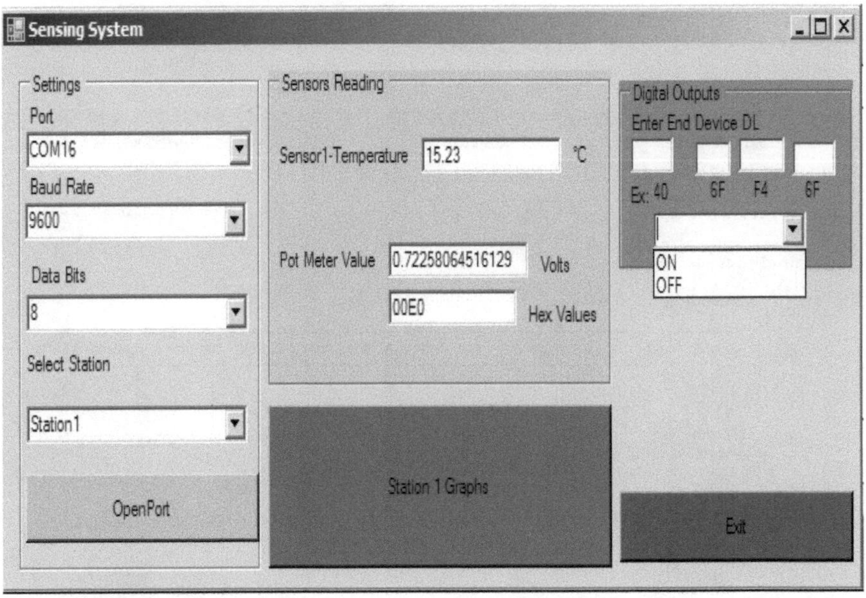

Fig. 5.11 GUI of the Sensor data reception from the sensing device #1

5.4 Testing the Communication between Coordinator and Remote XBee

In this section two examples are presented to show the data communication between the coordinator and the sensor-node.

5.4.1 Example 1

How to turn-on a digital output on a remote sensing device:

We use remote AT command request in API. The AT command is sent to the D0(Pin20).

The complete command to turn-on DIO 0 is

7E 00 10 17 07 00 13 A2 00 40 4B ED 83 FF FE 02 44 30 05 BB

0x44, 0x30 is for D0 and 0x05 is the parameter value.

For DIO1 Set the value to 0x31, DIO2 Set the value to 0x32 and so on. The X-CTU windows for the coordinator (transmitting the command) and the response from the sensor-node (received command) on the computer screen are shown in figures 5.12 and 5.13 respectively.

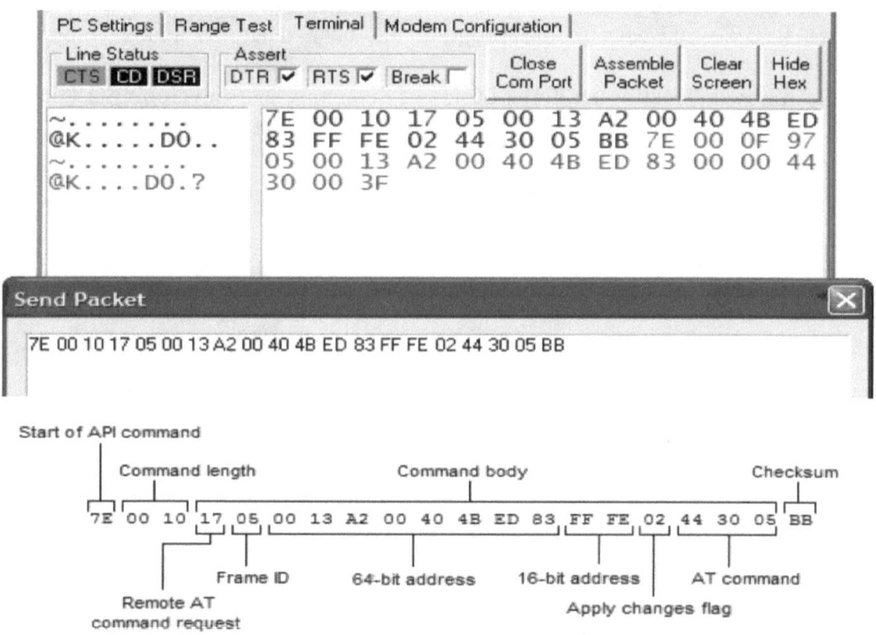

Fig. 5.12 The X-CTU window for sending the command to the sensor-node

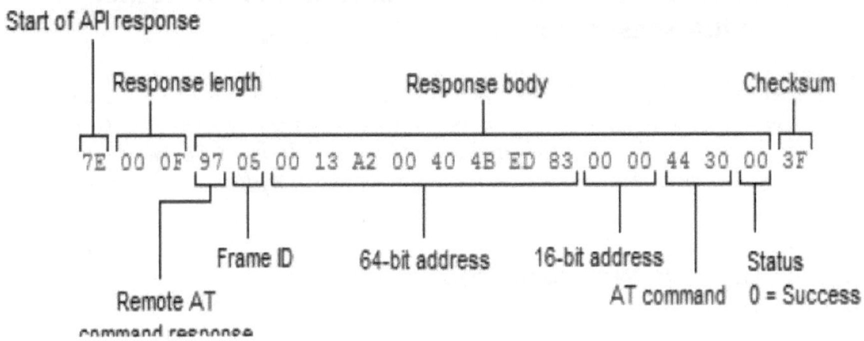

Fig. 5.13 The X-CTU window of the received response from the sensor-node

Example 2

How to Poll a remote XBee Module: The example as shown in figure 5.14 demonstrated how to poll a remote XBee module. It assumes that DIO0 is set as an Analog Input.

The API command to poll a remote XBee module is:

```
7E 00 0F 17 05 00 13 A2 00 40 4B ED 83 FF FE 02 49 53 98
```

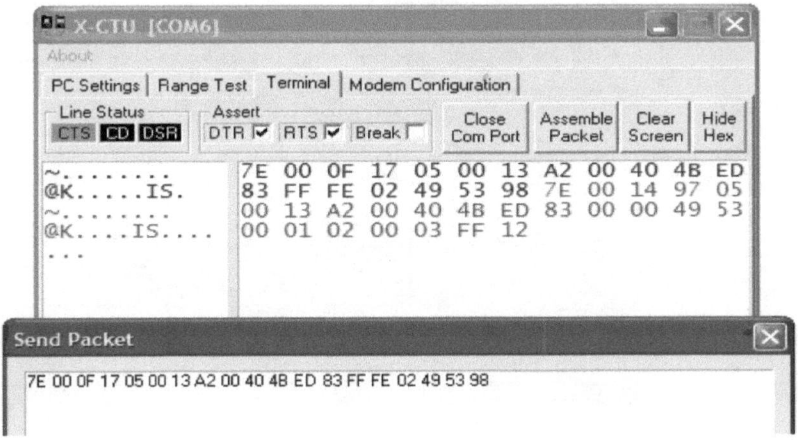

Fig. 5.14 The X-CTU window to poll a remote sensor-node

5.5 Design and Development of Graphical User Interface for Receiving Sensor Data Using C#

Note: Microsoft Visual Studio- C# programming language has been used for developing the User Interface. Visual Studio has the option of downloading and installing the free product for a free 90-day trial of Visual Studio Professional. It is to be made sure while downloading, C# support is enabled. Once Visual Studio is installed, it is launched from the Windows Start menu.

5.5.1 Creating a New Visual Studio C# Program

The following steps are followed to design a GUI (Graphical user Interface) for receiving the data by the computer and carry out necessary operations on them.

Step.1: Select a New Project from the File menu of Visual Studio.
Step.2: A New Project window appears containing a range of different types of project. For developing an user interface for receiving the sensor data, we will be developing a Windows Forms Application.
Step.3: Provide a File name as "***Sensingsystem_demo***" at the bottom of the dialog window and press the OK button to initiate the creation of the new project as shown in figure 5.15.

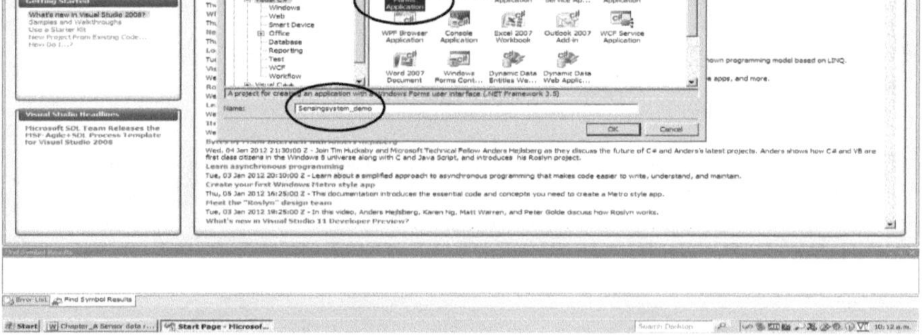

Fig. 5.15 Selecting C# windows forms application (Step.3)

Step 4: Once the new project has been created the main Visual Studio window will appear. At the center of this window there will be a new form in which we will create the user interface for our sensing system using C# application, as shown in figure 5.16.

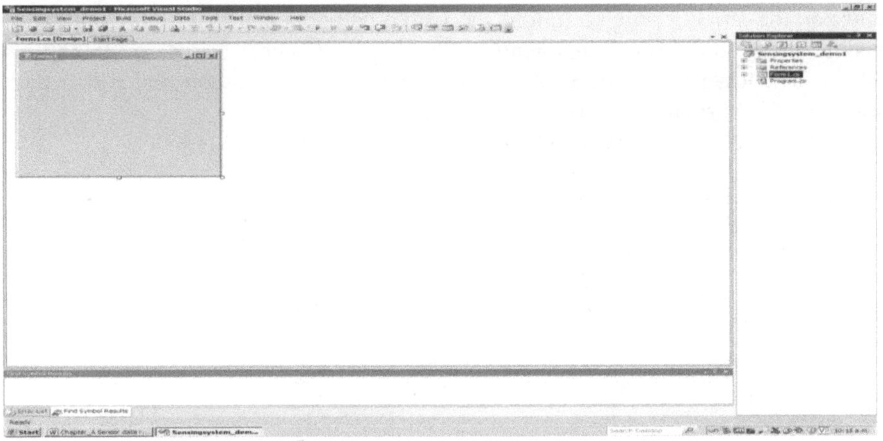

Fig. 5.16 Creating New form for adding components of GUI (Step. 4)

Step 5: Adding Components to the Windows Form

At this point we have a new Visual Studio project and are ready to begin the process of adding user interface components to our application. At the moment our Windows form (entitled Form1) is empty. The next step is to start dragging components from the Toolbox to the Form as shown in figure 5.17.

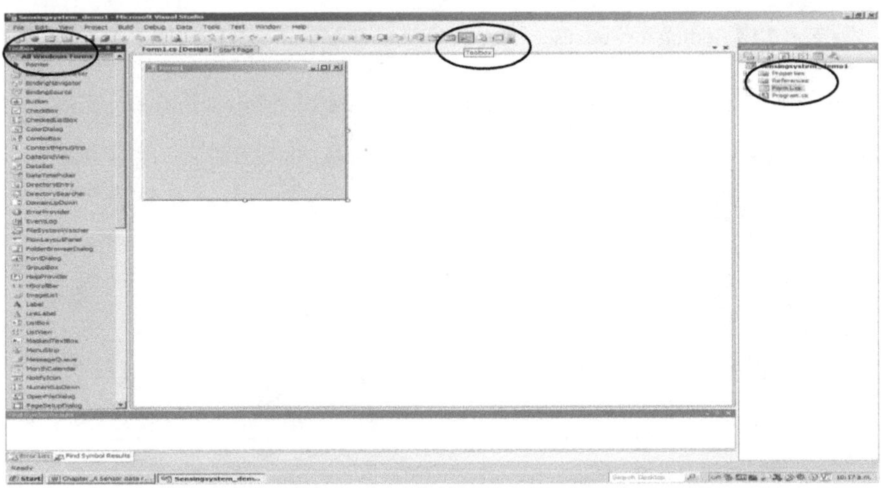

Fig. 5.17 Select components from the tool box and drag onto the form (Step 5)

To access the Toolbox click on the *Toolbox* tab located along the left hand side of the main Visual Studio window. This will display the Toolbox which contains a number of different categories of components available for addition to the Form as shown in figure 5.17. If the *All Common Components* category is currently folded click on the small + sign to unfold the list of components. With the components visible drag and drop components (Text box's, Combo boxes, push buttons, labels) as shown in figure 5.17 onto the Form canvas position and resize them such that the Form appears as shown in the figure 5.17.

Step 5(a): Add text boxes and labels them as shown in figure 5.18.

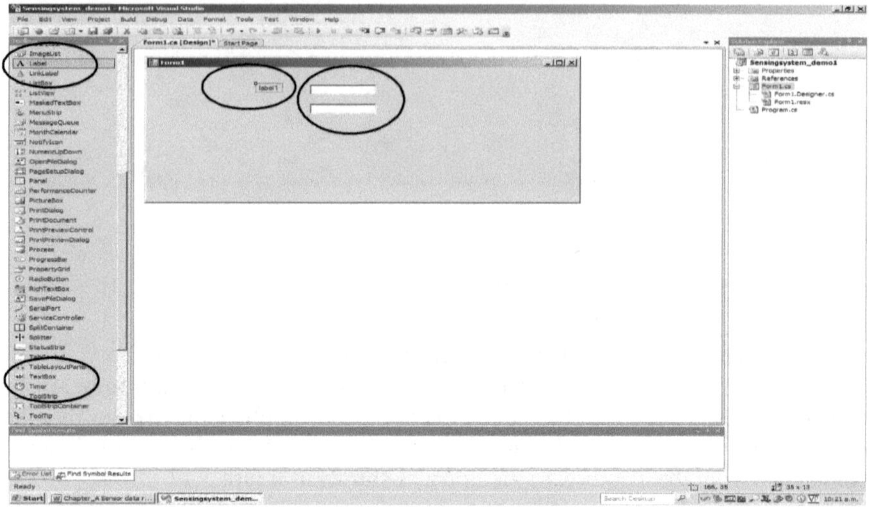

Fig. 5.18 Adding text boxes and labels by dragging from tool box (Step. 5(a))

Step 5(b): Change the properties of the components as shown in figure 5.19.

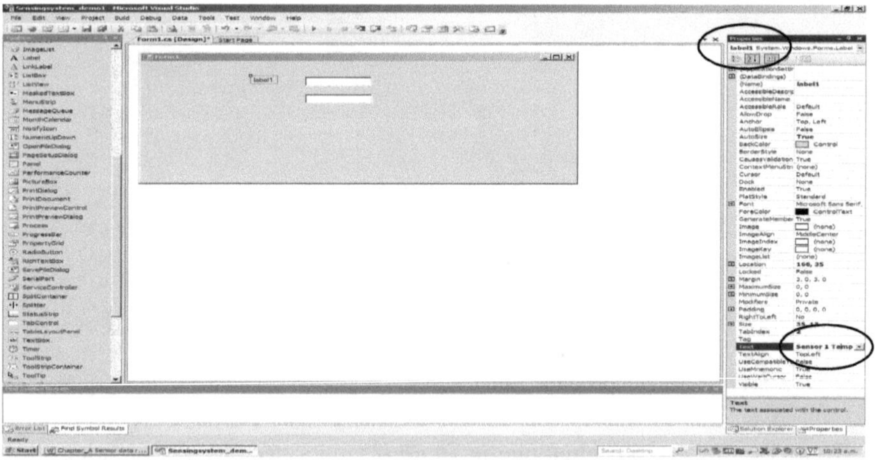

Fig. 5.19 Properties of the components are changed in the properties control (Step 5(b))

Step 5(c): Add combo boxes and change the properties as shown in figure 5.20.

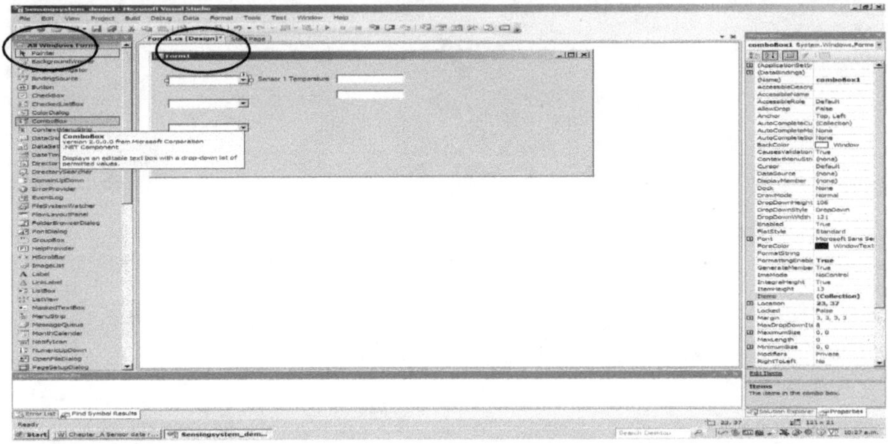

Fig. 5.20 Adding combo boxes and changing their properties (Step 5(c))

Step 5(d): Add serial port component as shown in figure 5.21.

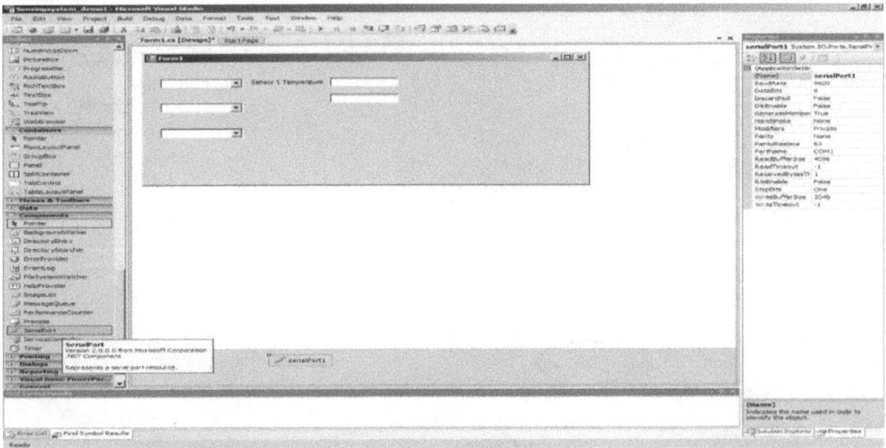

Fig. 5.21 Adding serial port component (Step 5(d))

Change the properties of the serial port component as shown in figure 5.22.

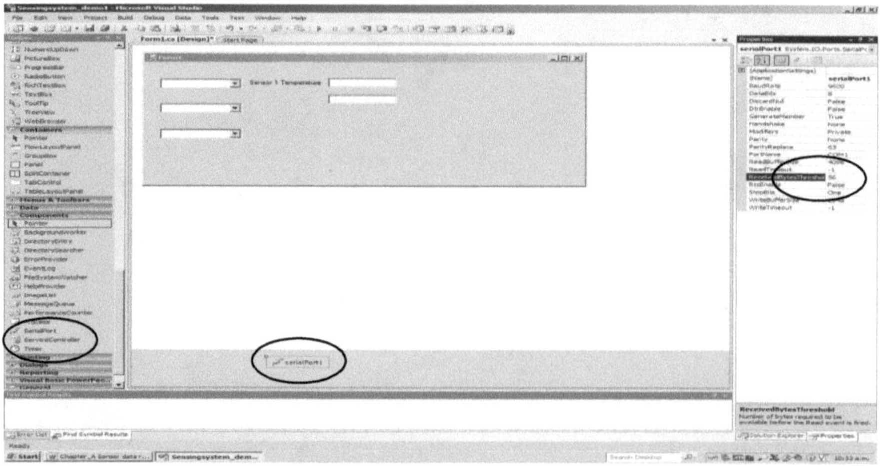

Fig. 5.22 Changing the properties of the serial port component

Step 5(e): Add timers to the GUI as shown in figure 5.23.

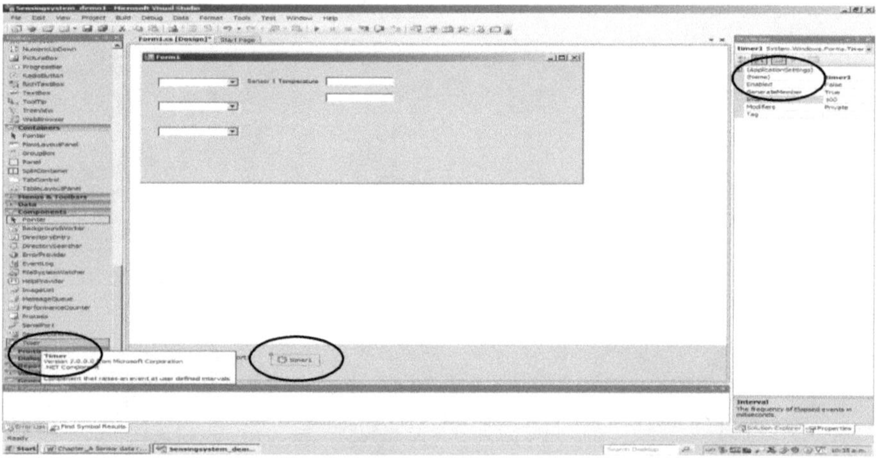

Fig. 5.23 Adding timer component to the GUI and changing the properties (Step 5(e))

Step 5(f): Add push buttons for openport, exit operations as shown in figure 5.24.

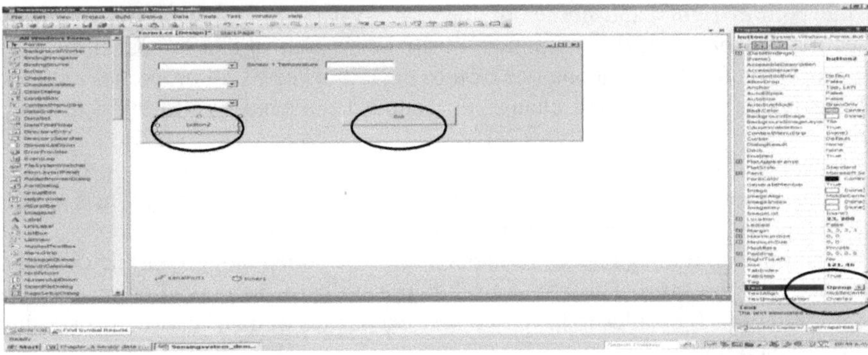

Fig. 5.24 Adding push buttons (Exit, Openport, .) and changing their properties(Step 5(f))

Complete the form by adding various components to the form as shown in figure 5.25:

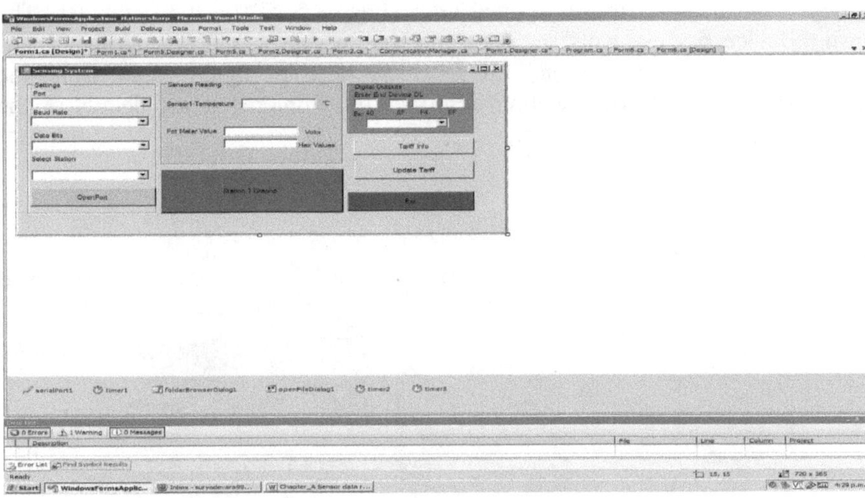

Fig. 5.25 Complete set of components added to the GUI

5.6 Changing Component Names

Visual Studio assigns default names to each and every component as and when they are added in to the Form. It is via these names that any C# code will interact with the user interface of the application. For this reason it is important to specify meaningful names which identify the component when referenced in the C#

source code. It is recommended, therefore, that the default names provided by Visual Studio be replaced by more meaningful ones. This and other properties relating to components are specified through the Properties panel which is located in the bottom right hand corner of the main Visual Studio window. Begin by selecting the top TextEdit component in the Form area so that the panel displays the properties for this component. Scroll to the top of the list of properties until the (Name) value is visible and change this name from textBox1 to temp_textbox, etc.,:

Repeat the above task by selecting each component in the Form in turn and changing the (Name) property. The second textBox should be named textbox_potval, the buttons openport and exitButton respectively.

"One important component "serialport" is to be added along with the common controls in to the form. This serialport component helps us to read or write to the USB serial port. It enables to have two way exchanges of data between the sensing device and computer."

The following section will explore data reception and transmission by the sensing device #1 and explore some ways of handling different tasks for data reception.

Figure 5.26 shows the coordinator (Receiving sensor data) on the X-CTU terminal interface and the developed GUI (Figure 5.27) on the computer. The sensing device has elements such as temperature sensor and potentiometer (for variable voltage readings) and an LED to receive digital input from the coordinator. The following section describes the implementation details of the development of the GUI interface.

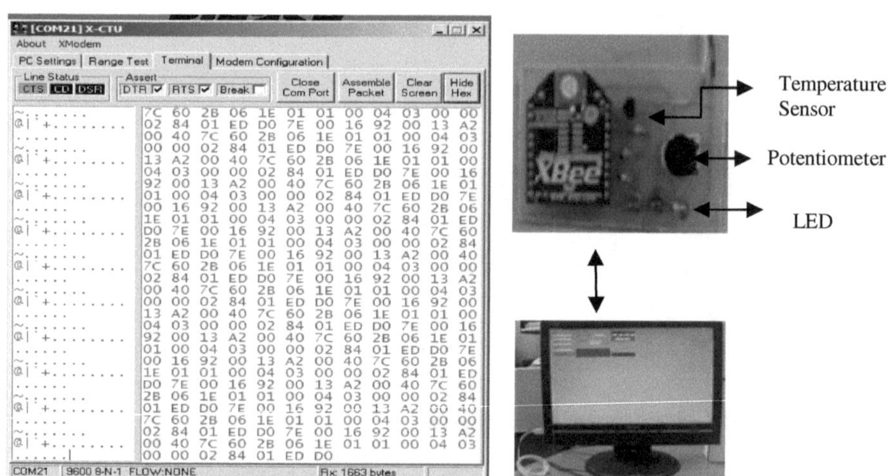

Fig. 5.26 X-CTU interface (Coordinator) receiving continuous data

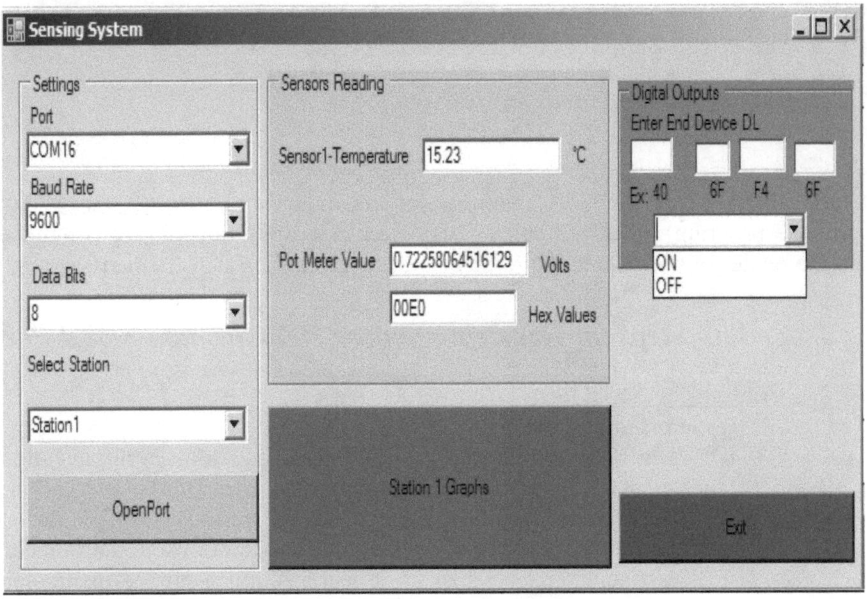

Fig. 5.27 GUI receiving sensor data

5.7 Add Program Statements to a Visual Studio C# Application

The next task in creating our application is to add some functionality so that things happen when we receive sensor data in our form. This behaviour is controlled via events. For example, when a button is pressed a Click event is triggered. All we need to do, therefore, is write some code for the Click events of our buttons.

Ex: When the exitButton is pressed by the user we want the application to exit. We can achieve this by calling the exit() method in the exitButton_Click event:

```
private void exitButton_Click(object sender, EventArgs e)
        {
            Exit ();
        }
```

The next task is to read the sensor data from the USB serial (coordinator) and use it to display and store in computer for further processing. This will take place when *openport* button is pressed. Click on the Form1.cs [Design] tab above the code area to return to the visual view of the Form and double click on openport to access the code. This time we will be adding code to the openport_ButtonClick() event method to read the values from USB SERIAL PORT. Once the coding is completed, build and run the application by pressing F5. Press the exit button to close the application.

Ex:

```
private void Openport_ButtonClick(object sender, EventArgs e)
{
serialPort1.DataReceived+=new
                SerialDataReceivedEventHandler(serialdata);
}
```

Details about event methods required for receiving and storing sensor data (complete program) are given below. Statements starting with "//" are comments help the reader to understand the logic of the corresponding steps in the program.

5.7.1 Coding Steps for Receiving Sensor Data through Serial Port

1. Configuring Serial Port:
   i) *Portname:* The name of the serial port.
   ii) *BaudRate:* The value of Baud Rate of serial communication. Select 9600 bps which is standard data rate.
   iii) *Databits:* Defines how many data bits are to be received during communication. It's always a good choice to use 8 data bits, as it's a standard byte size, so it becomes much easy to manipulate it.
   iv) *StopBits:* The number of stop bits to be used. The possible values are 1, 2, or none. I have used None.

   Code in program:

   ```
   ..........
   serialPort1.PortName = cboPortnames.Text;
   serialPort1.BaudRate = cboBaudrates.Text;
   ..........
   ```

2. Receiving Data:
   After the Serial Port is opened by calling serialport.open() method, the serial port is ready for receiving the data arriving at it.
   DataReceived: To receive data at the application, we need to handle DataReceived event of SerialPort class.

   ```
   ..........
   serialPort1.DataReceived += new
   SerialDataReceivedEventHandler(SData);
   ..........
   private void SData(object sender,
   SerialDataReceivedEventArgs e)
   {this.Invoke(new EventHandler(DoUpdate));}
   private void DoUpdate(Object s,EventArgs e)
   {........} ..........
   ```

3. Displaying Data:
   i. Conversion from Hex values to Decimal.
   ii. Apply formulas to decimal values for displaying sensor data correctly.
4. Storage: Sensor data is stored in a file.

   ```
   ..........
   StreamWriter fileStream = new
   StreamWriter(@"c:\STATION1_SENSORDATA.txt", true);
   fileStream.Write(DateTime.Now);fileStream.Write(",");
   ..........
   ```

5.8 The Complete Program(*Form1.cs*) with Explanation in the Form of Comments is Given Below

```
using System;                                    //Header files
using System.Collections.Generic;
using System.ComponentModel;
using System.Data;
using System.Drawing;
using System.Linq;
using System.Text;
using System.Windows.Forms;
using System.IO.Ports;
using System.Timers;
using System.Data.OleDb;
using Sensingsystem_demo;
using System.IO;
namespace Sensingsystem_demo
{
   public partial class Sensingsystem : Form
   {
     // Sensingsystem class contain all the member methods and
     // variables required for receving sensor data and send
     // data to sensing device when required.
  long Adc02; // Variable to receive ADC values for channel '0'.
  long Adc12; // Variable to receive ADC values for channel '1'.
  long Addr;  // variable to hold the address of the sensing device.
  public double s1, s2, s3; // VARIABLES TO HOLD SENSOR DATA
  public string h1, h2, h3, h4, h5, h6;
  public Sensingsystem()
  {
Initializecomponent();
     //list of available computer portnames in the combobox
          List<String> tList = new List<String>();
          foreach (string s in SerialPort.GetPortNames())
           {
               tList.Add(s);
           }
          tList.Sort();
          cboPortnames.Items.Add("Select USB Serial port...");
          cboPortnames.Items.AddRange(tList.ToArray());
          cboPortnames.SelectedIndex = 0;
    }
```

Figure 5.28 shows the output while the above programme is executed; though it is not only for the above programme.

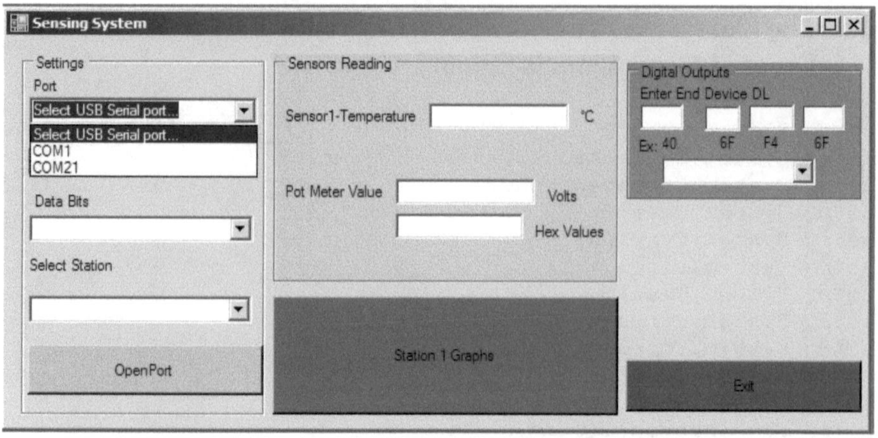

Fig. 5.28 Selecting the Coordinator USB serial port

```
private void Openport_ButtonClick(
                              object sender,EventArgs e)
{
//when openport button is pressed
serialPort1.PortName = cboPortnames.Text;//Serial port selection.
serialPort1.BaudRate = convert.ToInt32(cboBaudrates.Text);//Baud
Rate selection.
serialPort1.Open();           //Serialport open for receiving data
serialPort1.DataReceived += new
                 SerialDataReceivedEventHandler(SData);
                        // Serial data receving.
 }
 private void SData(object sender,
                 SerialDataReceivedEventArgs e)
 {
 // separate thread to handle receiving of serial data continuously.
 this.Invoke(new EventHandler(DoUpdate));
 }
```

Figure 5.29 shows the output of the baud rate due to the execution of the above programme. After selecting the port and the Baud rate, the open port is pressed for selection.

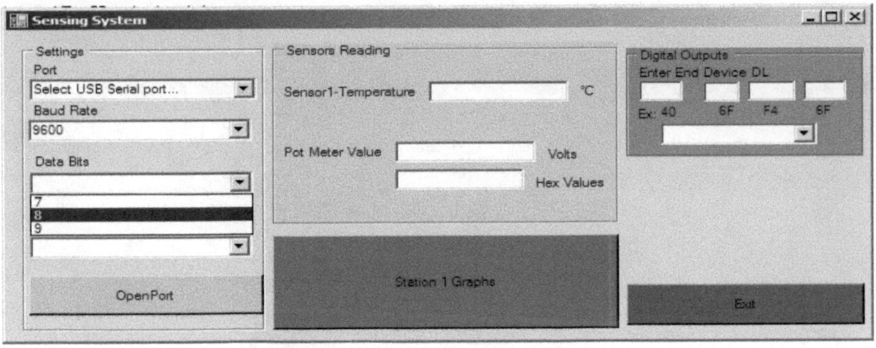

Fig. 5.29 USB serial Port settings

```
private void DoUpdate(Object s,EventArgs e)
{
//Temporary buffer to hold sensor  bytes of data.
//Since we used 2 channels for receiving analog values and 1
channel to
//send digital data we need 52 bytes of data packet to exchange
with
//sensing device.
   byte[] rcbuf = new byte[52];
   serialPort1.Read(rcbuf, 0, 51);
  //serial port reading 52 bytes of data at once into temporary
buffer.
   string hex = null;
   hex = ByteToHex(rcbuf);// to hold hex value.
   //
   if (hex.StartsWith("7E"))//starting value of packet.
   {
     hex = hex.Replace(" ", "");// remove spaces between hex
values.
     byte[] raw = new byte[(hex.Length / 2)];
          int i = 0;
          for (i = 0; i <= raw.Length - 1; i++)
{
//convert hex value
raw[i] = Convert.ToByte(hex.Substring((i * 2), 2), 0x10);
}
//address of sensing device is in 18 and 19 bytes of packet.
Addr= Int32.Parse(hex[18].ToString() + hex[19].ToString()
             System.Globalization.NumberStyles.HexNumber);
//if the user selects station1 to monitor
if (cboSensor.SelectedItem == "Station1")
{
// get the channel_0 sensor data into variable ADC02
Adc02 = Int32.Parse(hex[42].ToString() + hex[43].ToString() +
hex[44].ToString() + hex[45].ToString(),
System.Globalization.NumberStyles.HexNumber);
// get the channel_1 sensor data into variable ADC12
Adc12 = Int32.Parse(hex[46].ToString() + hex[47].ToString() +
hex[48].ToString() + hex[49].ToString(),
System.Globalization.NumberStyles.HexNumber);
```

The corresponding output is shown in figure 5.30.

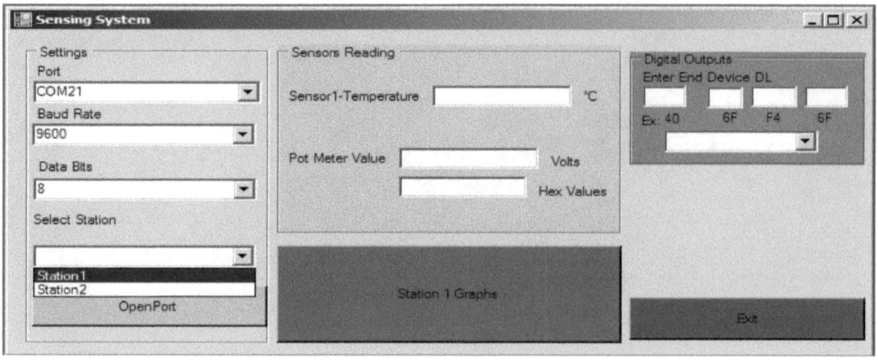

Fig. 5.30 Selecting the sensing device to receive the sensor data

```
//OBTAINING THE EQUIVALENT TEMPERATURE VALUE FROM THE
   //SENSOR DATA(ANALOG) BY APPLYING FORMULA OF THE
   //SPECIFIC TEMPERATURE SENSOR AND STORING IN THE
   //VARIABLE S1.
                    if (Adc02 > 0)
                     {
                         s1 = (3.3 * Adc02) / 1023;
                         s1 = (s1 * 1000 - 500) / 100;
                      }
  //OBTAINING THE EQUIVALENT POT METER READING FROM THE
  //ANALOG VALUE BY APPLYING FORMULA AND STORING
  //IN THE VARIABLE S2.

                   if (Adc12 > 0)
                     {
                         s2 = (3.3 * Adc12) / 1023;
                     }
                }
             }
            temp_TXTBOX.Clear();
            pot_TXTBOX.Clear();
if (cboSensor.SelectedItem == "Station1")
  {
   //DISPLAYING SENSOR DATA IN THE TEXT BOXES.
      String ss1 = string.Format("{0:N2}", s1);
  temp_TXTBOX.Text = ss1.ToString();
  pot_TXTBOX.Text = ((3.3 * Adc12) / 1023).ToString();
//Store sensor data in a text file(name of the file
and location should be specified)
```

```
//To write data in a file we can use Streamwriter
class of C#.
StreamWriter fileStream = new
                StreamWriter(@"c:\file100.txt", true);
                 fileStream.Write(DateTime.Now);
                 fileStream.Write(",");
                 fileStream.Write(temp_TXTBOX.Text);
                 fileStream.Write(",");

fileStream.WriteLine(pot_TXTBOX.Text);
                 fileStream.Close();
                }
          }

void timer_Tick(object sender, EventArgs e)
//To display sensor data continuously we can use timer
//function to refresh sensor data text boxes regularly.
     {
              temp_TXTBOX.Clear();
               pot_TXTBOX.Clear();
      if (cboSensor.SelectedItem == "Station1")
       {
       Temp_TXTBOX.Text = s1.ToString();
       Pot_TXTBOX.Text = ((3.3*Adc12)/1023).ToString();
       }

     }
 private string ByteToHex(byte[] comByte)
 {
//Function to convert hex values.
StringBuilder builder = new StringBuilder(comByte.Length * 3);
foreach (byte data in comByte)
builder.Append(Convert.ToString(data, 16).PadLeft(2,
'0').PadRight(3, ' '));
return builder.ToString().ToUpper();
}

private void exitbutton_Click(object sender,EventArgs e)
          {
               serialPort1.Close();
               this.Close();
          }
```

The corresponding output of the programme is shown in Figure 5.31.

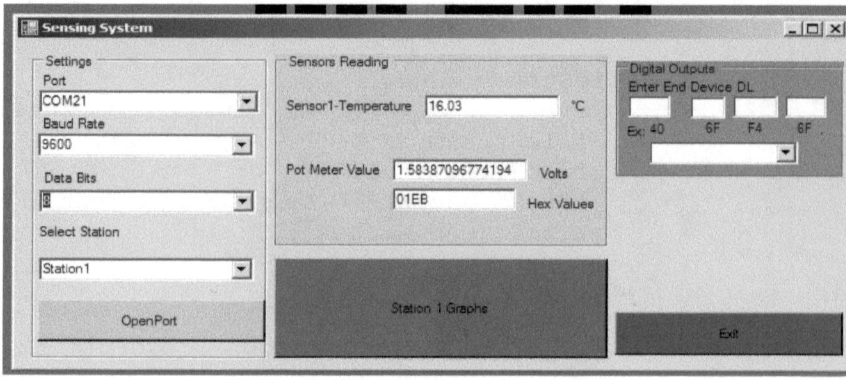

Fig. 5.31 Selecting Exit on the GUI

```
public void on_option(String ynn)
{
//If user want to switch on the LED on the sensing
device.
if (ynn == "ON")//on_option();
  {
//Packet to send digital high(ON)to the pin 17 of
sensing //device.
byte[] mhex = { 0x7E, 0x00, 0x10, 0x17, 0x05, 0x00,
0x13, 0xA2, 0x00, 0x40, 0x8A, 0x7A, 0x2A, 0xFF, 0xFE,
0x02, 0x44, 0x32, 0x05, 0x44 };
//WRITE a command to the USB serialport to send
digital //value to the sensing device.
    SerialPort1.Write(mhex, 0, mhex.Length);
    }
 }
```

The corresponding output of the programme is shown in Figure 5.32.

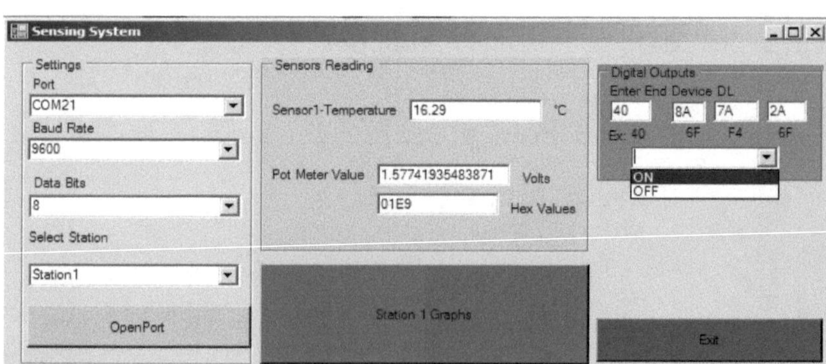

Fig. 5.32 Specifying Address of sensing device and selecting ON OPTION FOR LED

```
public void off_option(String ynn)
{
//Packet to send digital LOW(OFF)to the pin 17 of
sensing //device.
            if (ynn=="OFF")
           {
byte[] mhex = { 0x7E, 0x00, 0x10, 0x17, 0x05, 0x00,
0x13, 0xA2, 0x00, 0x40, 0x8A, 0x7A, 0x2A, 0xFF, 0xFE,
0x02, 0x44, 0x32, 0x04, 0x45 };
             serialPort1.Write(mhex, 0, mhex.Length);
           }
         }
}
```

The corresponding output of the programme is shown in Figure 5.33.

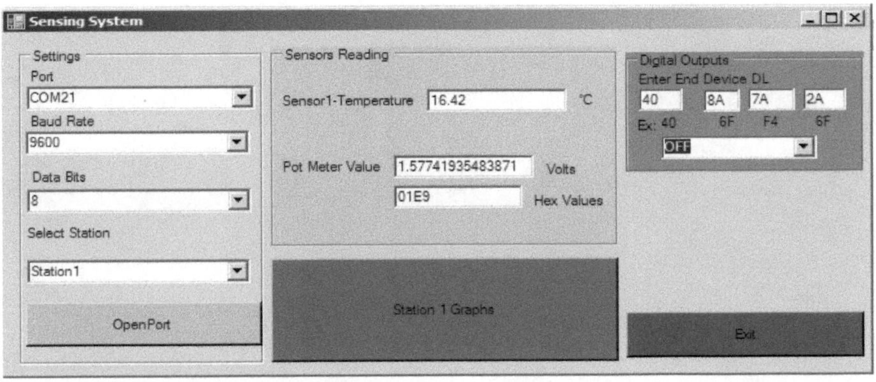

Fig. 5.33 Specifying Address of sensing device and selecting OFF OPTION FOR LED

Though the above programme is an elementary programme developed for the data reception, storage and analysis, this can be extended to make it more complicated to suit the requirements.

References

1. http://en.wikipedia.org/wiki/ZigBee
2. http://www.digi.com/products/wireless-wired-embedded-solutions/zigbee-rf-modules/point-multipoint-rfmodules/xbee-series1-module#specs
3. http://www.sensorsmag.com/networking-communications/wireless-sensor-network-topologies-778
4. http://code.google.com/p/xbee-api/wiki/WhyApiMode
5. http://www.digi.com/support/kbase/kbaseresultdetl?id=2206
6. http://www.digi.com/support/kbase/kbaseresultdetl?id=3221

Chapter 6
Sensors Signal Processing Techniques

Introduction

This chapter provides an overview of a few signal processing techniques related to sensing phenomenon. Though there are so many techniques used and/or is available, it is practically impossible to describe each and every one in this chapter. Moreover, the Matlab provides a wide range of different methods in the Signal Processing Toolbox which the readers may go through. The raw signal available from the sensors are usually passed though a few hardware circuits depending on the applications. The purpose is to eliminate undesired noises presence in the signals as well as making the signal strong through a preamplifier stage. In this chapter a few method will be described which are used by the author in the research.

6.1 A Brief Review of Signal Processing Techniques for Structural Health Monitoring

The sensing system for monitoring the health of structures is becoming an increasing important topic of research in recent times. Different methods of new sensors and sensing systems are reported. The sensors raw data are analyzed to derive the useful information. The sensors in a wireless sensor networks are distributed over a large area. The raw data can be collected from the sensors through wireless communication and all data can be gathered in the central processor situated at a far distant place. For the transmission of data, a comparison has been made between a single-hop data transmission and multi-hop data transmission. It has been shown that the throughput of a single source node decreases as the number of hops increases. This is due to the interference from 2- or 3-hop nodes, even though the radio range is only one hop.

One of the requirements of employing WSN for some applications is that the amount of wireless communication required by the algorithm to be as minimum as possible to save energy and decrease packet loss rate. The sensor node should be light weight. Moreover, the resource consuming algorithms are not suitable to be implemented in smart sensor nodes. The whole software design should be such that it should be able to accommodate energy saving strategies like wakeup and sleep scheduling. The algorithm should be developed to minimize false-positive

S.C. Mukhopadhyay: *Intelligent Sensing, Instrumentation and Measurements*, SSMI 5, pp. 119–139.
DOI: 10.1007/978-3-642-37027-4_6

and false-negative indication of any event. The software should be able to detect any important or undesirable development at an early stage and should also provide the location of the incident. A combination of ACF (Auto-correlation function) and CCF (Cross-correlation function) can be useful for many applications. For example, for structural health monitoring (SHM) applications, if there is any damage in the structure, the ACF of the obtained time series will be different from those obtained from the undamaged structure. It is expected that the ACF is sensitive to damage but not to input/environmental changes. If the ACF algorithm provides indication of damage then the CCF is used to locate the damage. The sensor nodes deployed on the structure are divided as node pairs. Each node pair covers an area of the structure where the node pair is located. If damage occurs in that area, the dynamic relationship of the two nodes in the node pair will be changed. The CCF of the two nodes will be altered correspondingly.

A considerable effort in signal de-multiplexing is required to obtain correct information towards realizing a practical system for SHM applications based on optical fiber based strain sensors. The strain measurements were made separately recording the wavelength of the maxima in reflection from each Bragg grating. Two forms of electronically tunable optical filters were developed as de-multiplexing systems.

A very useful method in SHM analysis is distributed processing – network intelligence. The intelligent sensing network must produce intelligent responses to the sensed environment and it should be on-line. The response should be produced by self-organization: it is emergent behavior of the system. The SHM problem involves multiple inter-related hierarchical sub-tasks (e.g. damage detection, evaluation, diagnosis, prognosis and repair). The approach assumes that single cells may make fast and automatic responses to critical emergencies, while collection of cells may solve more complex hierarchical tasks including:

(i) Self-calibration and discrimination between component and sensor failures;
(ii) Formation of a dynamic artificial neural network, characterizing the nature of possible damage and producing a self-organizing diagnosis;
(iii) Self-scheduling of secondary inspections, maintenance or corrective actions based on information from the network while issuing warnings;
(iv) Direction of recovery resources, human or robotic, to the repair site.

Determining desirable and quantitative information from the raw data observed by SHM may be equivalent to solving an inverse problem. Optimization procedures are often used to solve such inverse problem because optimum values are easier to obtain the exact values and good enough for practical use.

Ant Colony Optimization (ACO) algorithms have been used in many applications including SHM. The algorithms use the ability of the agents to interact indirectly through changes in their environment by depositing pheromones and forming a pheromone trail. A form of autocatalytic behavior – allelomimesis, the probability with which an ant chooses a trail increases the number of ants that choose the same path in the past has also been employed. In the algorithm, ants are implemented as communication packets, policies are implemented via appropriate message passing, and cells are responsible for interpreting received packets or sending packets. Since ants cannot move into the cells with broken communication links, they are supposed to find the shortest paths around them using positively reinforced pheromone trails. In general, the ACO approach attempts to solve an optimization problem by iterating the following two steps:

(i) Candidate solutions are constructed in a probabilistic way by using a probability distribution over the search space,
(ii) Candidate solutions are used to modify the probability distribution in a way that is deemed to bias future sampling towards high quality solutions.

It is not easy to measure damage directly from the sensors' data obtained in SHM. Many parameters must be measured throughout the structure and should be utilized to assess the health of the structure. It is important to know the type, number and placement of sensors on the structure, which are problem dependent. The following methods of analysis are quite common in relation to SHM:

6.1.1 Normalization

It is applied to account for changes in structural response due to environmental conditions or structural loads which are not associated with any structural damage. Some parameters, such as elastic modules of a structure, being temperature dependent may have a significant effect on the dynamic of the structure. Normalization is required to ensure that the changes in the dynamics of the structure are due to change in temperature or any other environmental parameters, and is not interpreted as damage. Gain normalization is utilized by dividing the response by its peak amplitude or standard deviation. Once the effects of environmental changes are compensated, the remaining changes in the measured response are a direct result of changes in the structural state. Normalization is closely related to calibration. It is defined as the transformation of sensor output to a nominal value based on a known input and specified environmental conditions. Calibration does not, however, account for any effects the environment may have on a structure's dynamics.

Some of the Data Normalization Methods for SHM are: Novelty detection, Regression analysis, Singular value decomposition, Factor analysis, Subspace-based identification method, Auto-associative neural network, Lamb-wave propagation method. In the following section one example on novelty detection method is presented.

Ex: Novelty Detection Method: Novelty detection first builds an internal representation of the system's normal condition, and then examines subsequent data to see if they significantly depart from a normal condition.

The discordancy of a candidate outlier is measured by the following squared Mahalanobis distance measure 'D'

$$D = (x - \overline{x})^T C (x - \overline{x})$$

Where x is a vector of the potential outlier; $\overline{x}$ and C are the mean vector and covariance matrix-inverse of the baseline system, respectively. Once the Mahalanobis distance is computed, it is checked against a threshold value. Note that this novelty detection addresses only the simplest level of damage identification, i.e. whether damage is present or not. Major advantage of this novelty detection is that statistical model building for damage classification is based only on data from the undamaged system.

Example: Let there be two Variables X and Variable Y, and that our two variables had the following characteristics:

Variable X: mean = 500, SD = 79.32, Variable Y: mean = 500, SD = 79.25

Variance/Covariance Matrix		
	X	Y
X	6291.55737	3754.32851
Y	3754.32851	6280.77066

If, in our single observation, X = 410 and Y = 400, we calculate the Mahalanobis distance for that single value as: $D = (x - \overline{x})^T C(x - \overline{x})$

$$(x - \overline{x}) = \begin{pmatrix} 410 & - & 500 \\ 400 & - & 500 \end{pmatrix} = \begin{pmatrix} -90 \\ -100 \end{pmatrix}$$

$$C = \begin{pmatrix} 6291.55737 & 3754.32851 \\ 3754.32851 & 6280.77066 \end{pmatrix}^{-1} = \begin{pmatrix} 0.00025 & -0.00015 \\ -0.00015 & 0.00025 \end{pmatrix}$$

$$\Rightarrow \quad D = \begin{pmatrix} -90 & -100 \end{pmatrix} * \begin{pmatrix} 0.00025 & -0.00015 \\ -0.00015 & 0.00025 \end{pmatrix} * \begin{pmatrix} -90 \\ -100 \end{pmatrix}$$

$$\Rightarrow \quad 1.825$$

Therefore, our single observation would have a distance of 1.825 standardized units from the mean.

6.1.2 Feature Extraction

It is the process of computing metrics from sensor signals that have the potential to discriminate among the structural states to be identified. Desirable features are ones that are responsive to the structural damage states, yet insensitive to other factors. In many situations, the features are generated from simulation analysis.

Some of the methods are: Linear feature extraction methods: Resonant frequencies, Frequency Response functions, dynamic flexibility, Auto Regression Moving Average (ARMA) family models, Canonical variate analysis, Non-linear features: Time frequency analysis, Hilbert transform, and Finite model updating.

Ex: Auto Regression Moving Average (ARMA) family models: Sohn and Farrar (2000) have applied statistical process control techniques to vibration-based damage diagnosis. First, an Auto-Regressive (AR) mode is fit to the measured acceleration time histories from an undamaged structure,

$$y(t) = \sum_{j=1}^{3} \varphi_j y(t-j) + e(t)$$

where y(t) is the measured time history at time t, φ_j is the AR coefficients to be estimated, and e(t) is the prediction error term. The coefficients of the AR model are selected as the damage sensitive features for the subsequent control chart analysis. Then, the AR coefficients obtained from subsequent new data are monitored relative to the baseline AR coefficients. Any significant deviation from the baseline of AR coefficients would indicate either a change in environmental conditions or damage.

Assuming the total duration of the record of y(t) is 480 sec. The record is divided into 80 segments, denoted by $y_j(t)$, j=1,2,....80, each having 6 seconds duration sampled at 1000 Hz resulting in 6000 data points per segment. The AR coefficients are computed for each six second segment of the acceleration data. To determine the sensitivity of the coefficients to the number of data points in the signal, analyses were performed in the range of 1000 to 6000 points in increments of 1000.The AR coefficients were found to reach stable values at about 3000 points. However, 6000 points were used in the analysis. Figures 6.1 to 6.4 show the different related waveforms. The stability of the first AR coefficients with the number of data points is given in table 6.1. Both the mean and standard deviation of the coefficients are given.

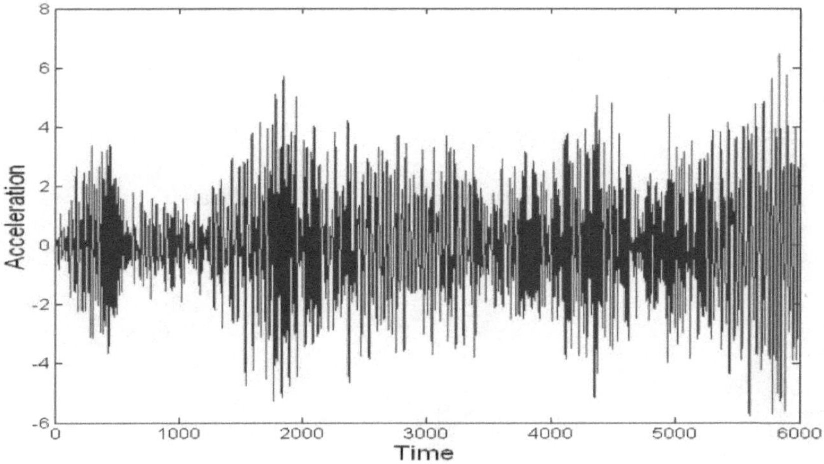

Fig. 6.1 A typical vibration signal from a sensor showing the model characteristics

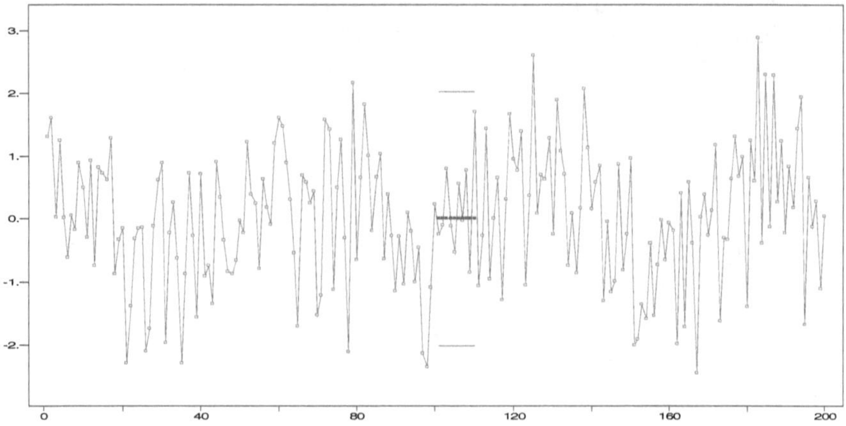

Fig. 6.2 Corresponding time series of the vibration signal sensor with 200 points

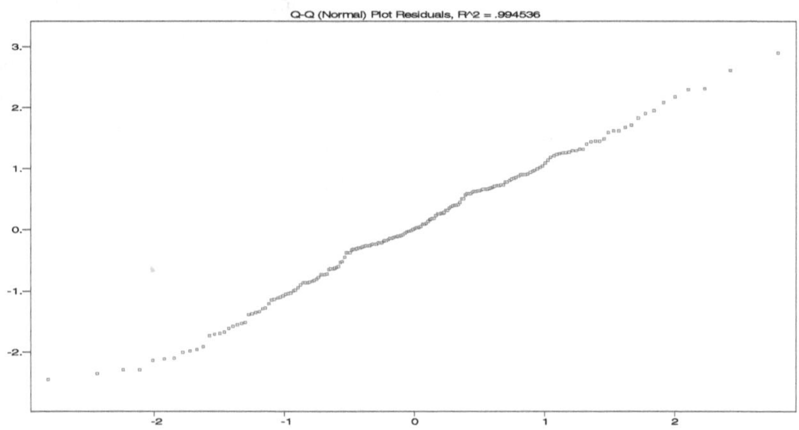

Fig. 6.3 Normal probability plot of the residuals

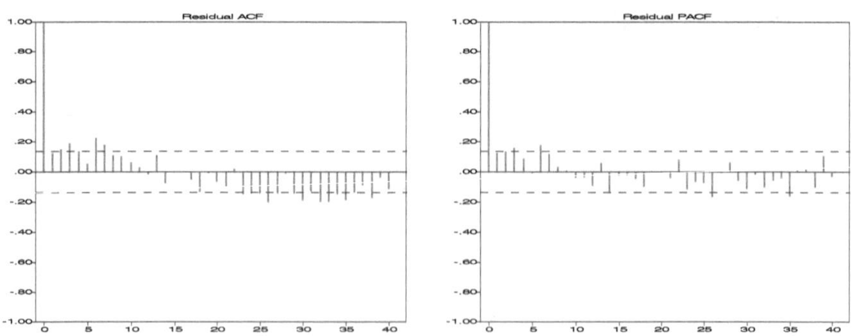

Fig. 6.4 Variation of auto correlation of the residuals with lag

Table 6.1 Sensitivity of AR coefficients with the number of data points

Value of AR Coefficients	Number of Data Points					
	1000	2000	3000	4000	5000	6000
Mean of α_1 (Std. Deviation of α_1)	1.0441 (0.1947)	1.0587 (0.1369)	1.0566 (0.1088)	1.0453 (0.0981)	1.0359 (0.0831)	1.0301 (0.0788)
Mean of α_2 (Std. Deviation of α_2)	1.0359 (0.1966)	1.0502 (0.1373)	1.0517 (0.1002)	1.0459 (0.0840)	1.0403 (0.0710)	1.0358 (0.0582)
Mean of α_3 (Std. Deviation of α_3)	1.2204 (0.1366)	1.2644 (0.0861)	1.2772 (0.0608)	1.2762 (0.0451)	1.2761 (0.0360)	1.2712 (0.0338)

6.1.3 Dimensionality Reduction

To simplify the problem, only a selected number of features are used. The feature selection is the process of finding a subset of the original features. It can be categorized into filter methods and wrapper methods. In a filter method, relevant features are determined solely based on attributes computed from the data. But, for the wrapper approaches, it is determined based on how well a subset of features performs when used in a classifier. Another method of dimensional reduction is to project the original feature space into a lower dimensional space. Principal Component Analysis (PCA) is commonly used for that. It is the optimum projection in the sense of capturing the maximum data variance for any specified number of basis vectors.

PCA algorithm

1. X ⬅ Create N x d data matrix, with one row vector x_n per data point
2. X ⬅ subtract mean x from each row vector x_n in X
3. Σ ⬅ covariance matrix of X
4. Find eigenvectors and eigenvalues of Σ
5. PC's ⬅ the M eigenvectors with largest eigenvalues

Step 1: Assume data is set of d-dimensional vectors, where nth vector is

$$x^n = \langle x_1^n \ldots\ldots x_d^n \rangle$$

Example: Data

X	2.5	0.5	2.2	1.9	3.1	2.3	2	1	1.5	1.1
Y	2.4	0.7	2.9	2.2	3.0	2.7	1.6	1.1	1.6	0.9

We can represent these in terms of any 'd' orthogonal basis vectors

By Finding $\langle u_1 \ldots\ldots u_m \rangle$ that minimizes $E_M \approx \sum_{n=1}^{N} \|x^n - \hat{x}^n\|^2$

Where $\hat{x}^n = \bar{x} + \sum_{i=1}^{M} z_i^n u_i$; where $\bar{x} = \frac{1}{N} \sum_{n=1}^{N} x^n$(Mean)

Figure 6.5 shows the plot of the data with mean subtracted.

Step 2: Zero Mean of the above data:

X	.69	-1.31	.39	.09	1.29	.49	.19	-.81	-.31	-.71
Y	.49	-1.21	.99	.29	1.09	.79	-.31	-.81	-.31	-1.01

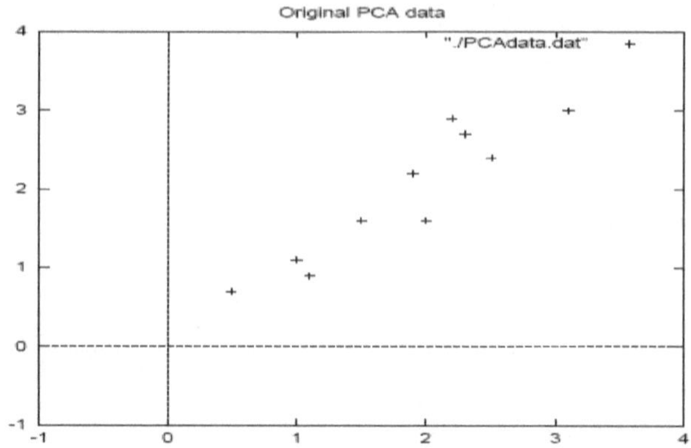

Fig. 6.5 Plot of PCA Example data with means subtracted

Step 3: Calculate the Covariance matrix:
Covariance Matrix: $\Sigma = \sum_n (x^n - \overline{x})\ (x^n - \overline{x})^T$

cov = $\begin{bmatrix} .616555556 & .615444444 \\ .615444444 & .716555556 \end{bmatrix}$

Since the non-diagonal elements in this covariance matrix are positive, we should expect that both the x and y variable increase together.

Step 4: Finding the eigenvector and eigenvalues:

Minimize $E_M = \sum_{i=M+1}^{d} u_i^T \Sigma u_i$

⇨ $\Sigma u_i = \lambda_i u_i$ ← Eigenvector
↑ Eigenvalue

⇨ $E_M = \sum_{i=M+1}^{d} \lambda_i$

eigenvalues = $\begin{bmatrix} .0490833989 \\ 1.28402771 \end{bmatrix}$ and

eigenvectors = $\begin{bmatrix} -.735178656 & -.677873399 \\ .677873399 & -.735178656 \end{bmatrix}$

Figure 6.6 shows a plot of the normalized data (mean subtracted) with eigenvectors of the covariance matrix over-layed on top.

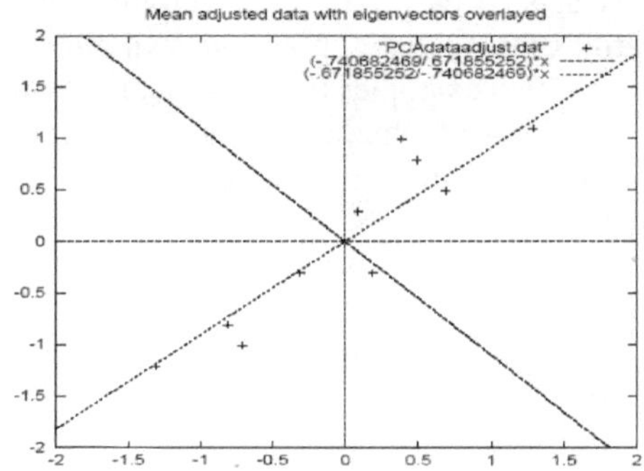

Fig. 6.6 A plot of the normalized data (mean subtracted) with eigenvectors of the covariance matrix over-layed on top

6.1.4 Collaborative Damage Event Detection (CBED) Method

This method is quite useful in SHM. In this method, each note continuously collects measured responses and checks the existence of any abrupt changes, which is assumed to be the indication of damage. In the case of abrupt responses, it communicates with neighbors to confirm the existence of damage. Usually damage on structures is accumulated slowly and doesn't incur abrupt changes. It is also difficult to distinguish the change caused by the real damage from the one caused by environmental conditions (noise or input change) and is prone to giving false positive alarms.

6.2 Signal Processing Techniques for Information Extraction from Sensor Data

In this section, few techniques which are used in a smart home project for analyzing the sensors data are presented. Though the methods are used for analysis in a smart home environment project, most of the methods are applicable in different projects too. The raw data obtained from the different sensors carry the useful information, but that information needs to be extracted. In order to extract the correct information a range of different signal processing techniques are used. Importance of these techniques is related to applicability of various signal processing methods on heterogeneous data fusion with information extraction.

In general, the healthy lifestyle of an elderly can be monitored by the Activities of Daily Living (ADL's) such as dressing, drinking, eating, cooking etc.,. The first step towards the development of efficient wellness model of an elderly in smart home is recognition of daily activities of the individual accurately.

In a survey for assistive technological innovation it is appeared that the recognition and engagement in an activity by Alzheimer's disease patients is rated as the highest by family care providers. Therefore researchers put forward different techniques to form and recognize daily activities. It is believed that each individual continuously acts on each individual activities of ADL in a pre-defined strategy at domestic atmosphere having more easily of guidance, which is contrary to the existing conditions. For example: if regular intake of food is regarded as an ADL activity, it cannot create out whether the individual is going out to have food or is having food at house which creates the tracking of ADL activity quite testing. An activity implemented under careful atmosphere may differ with people carrying out activities in a different way. It creates the dependency in the catalog of pre-defined activities unrelated because of the difference in inter contents. Further just one activity can be implemented in a different way by individual personal needing approaches to pact with intra-subject inconsistency. A few methods will be described in the following section.

6.2.1 Deriving Information from Sensor Data: Daily Activity Recognition Models

There have been several machine learning strategies for daily activity recognition which varies considerably. For activity recognition Naïve Bayes classifiers have been used with forecasting results. The classifiers identify the action that associates with the biggest possibility to set of indicator information those are observed. Despite the fact that these classifiers take up the depending freedom of features, and also give good precision when provided with lots of example information. To learn important actions, several researchers have used decision trees. These methods offer the benefit of generating guidelines which are easily understood by the user, but it is frequently subtle, when to gather the high precision number information. There is another optionally available strategy which has been examined by other researchers is to set probabilistic series of indicator effects using powerful Bayes systems, Conditional Random Fields and Markov models.

The main objective of activity recognition is to recognize common individual activities in the actual life situation. The accurate recognition of activity is very challenging due to several reasons and diversities of individual actions. Many algorithms have been devised to build activity recognition prototypes. The most appropriate designs are The Hidden Markov Design and the Conditional Random Field.

6.2.1.1 The Hidden Markov Model (HMM)

Markov Models: Consider the scenario of "Classifying Activity". Let the activity on a particular day can be any one: Breakfast, Sleeping or Showering. A mathematical prototype can be modeled based on the statistics. If we can know on what the weather S_n is like today (on day n) depending on what the weather was like yesterday S_{n-1}, then by applying conditional probabilities

$$P\ (S_n|S_{n-1})$$

the probability of the unknown weather at day n, $S_n \in$ {Breakfast, Showering, Sleeping}, depend on the (known) weather S_{n-1} of the past day. This is known as first-order Markov assumption. In other terms, the probability of a certain observation at time n only depends on the observation S_{n-1} at time n − 1. A second-order Markov assumption would have the probability of an observation at time n depend on S_{n-1} and S_{n-2}. In general, a Markov assumption, usually mean the first-order Markov assumption.

In general, the probability of a certain sequence {S1, S2,. . . Sn} (the joint probability of certain past and current observations using the Markov assumption is given by:

$$P(S_1, \ldots, S_n) = \prod_{i=1}^{n} P(Si|Si-1)$$

Ex: Classify Elderly daily activities into three states–

State 1: Sleeping, State 2: Showering and State 3: Breakfast

By carefully examining the Elderly daily activities for a long time, we found following activity pattern as shown in the following table 6.2.

Table 6.2 Probability of different activities of an elderly

Elderly Activity			
Today	Tomorrow		
	Breakfast	Sleeping	Showering
Breakfast	0.8	0.05	0.15
Sleeping	0.2	0.6	0.2
Showering	0.2	0.3	0.5

Note: Tomorrow's weather depends on today's weather.

Given that, now (today at a particular time) the elderly is in Breakfast activity, what's the probability that the next day (at the same time) will be Breakfast and then the day after is Sleeping? Figure 6.7 shows the presentation of the transition probabilities using the Markov model.

Using the Markov assumption and the probabilities in table 6.2, this translates into:

P(q2 =Breakfast, q3 =Sleeping |q1 =Breakfast)
= P(q3 =Sleeping |q2 =Breakfast , q1 =Breakfast) · P(q2 =Breakfast |q1 =Breakfast)
= P(q3 =Sleeping |q2 =Breakfast) · P(q2 =Breakfast |q1 =Breakfast)
(Markov assumption)
= 0.05 · 0.8 => 0.04.

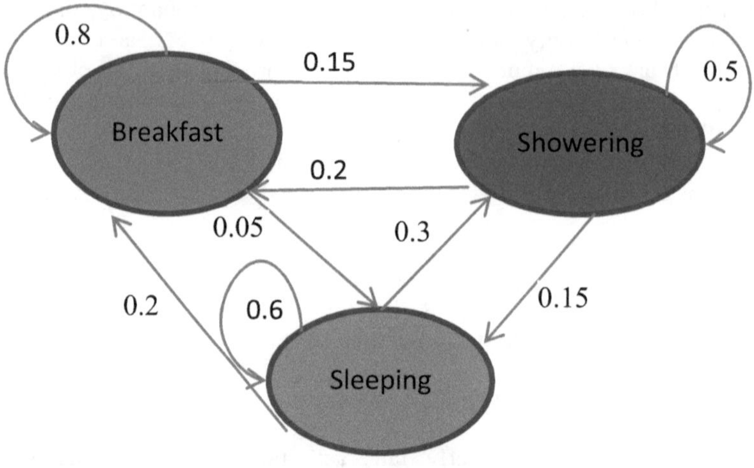

Fig. 6.7 Markov model with the transition probabilities according to table 6.2

Hidden Markov Models (HMMs)

In this case, the actual activity being performed is "*hidden*". It is assumed that the probability that elderly uses a particular object is 0.1 if he is having breakfast, 0.8 if it is actually sleeping, and 0.3 if it is showering.

Finding the probability of a certain activity q_i, ε {Breakfast, Sleeping, Showering} can only be based on the observation xi, with xi =1, if you get a response from the specific appliance on a day i, and xi =0 if you don't get a response. This conditional probability $P(q_i|x_i)$ can be rewritten according to Bayes' rule:

$$P(q_i|x_i) = \frac{P(x_i|q_i)P(q_i)}{P(x_i)} \text{ or}$$

for n days and activity sequence $Q=\{q_1,\ldots\ldots q_n\}$ as well as 'object sequence' $X=\{x_1,\ldots x_n\}$

$$=P(q1,\ldots\ldots,qn|x1,\ldots\ldots,xn) = \frac{P(x1,\ldots\ldots,xn|q1,\ldots\ldots qn)P(q1,\ldots\ldots qn)}{P(x1,\ldots\ldots,xn)}$$

We can draw conclusions from our observations (if the elderly performs an activity with the specific object or not). Therefore, omit the probability of seeing an specific object P(x1, . . . , xn) as it is independent of the elderly activity, that we like to predict. We get a measure for the probability, which is proportional to the probability, and which we will refer as the likelihood L.

$$P(q_1,\ldots,q_n|x_1,\ldots,x_n) \infty$$
$$L(q_1,\ldots,q_n|x_1,\ldots,x_n) = P(x_1,\ldots,x_n|q_1,\ldots,q_n) \cdot P(q_1,\ldots,q_n)$$

HMM is a generative probabilistic model implemented for creating hidden states from obvious data. Normal activities can be accurately formed in to Markov Stores. It is possible to create a model of action by tracking alerts from sophisticated and unidentified activities. These prototypes or designs are known as Hidden (Invisible) Markov Design or HMM. HMM research the effect of an action and gradually produces the action model keeping the likelihood of further adjusting, expansion and reusability for related research. The accurate purpose of HMM model is to recognize hidden state series (y1, y2, y3......) that suits with trial outcome series (x1, x2........., xt). It also tries to understand model aspects from the account of noticed outcome series.

A HMM model is specified by:

The set of states' $S=\{S_1,S_2,.....S_N\}$, and set of parameters $\Theta=\{\prod,A,B\}$:

--The *prior probabilities* $\prod_i = P(q_1=S_i)$ are the probabilities of Si being the first state of a state sequence, collected in vector $\prod$.
--The *transition probabilities* are the probabilities to go from state i to state j. They are collected in matrix A.
--The *emission probabilities* characterize the likelihood of a certain observation x, if the model is in state s_i, depending on the kind of observation x. The graphical representation of a hidden Markov model is shown in figure 6.8.

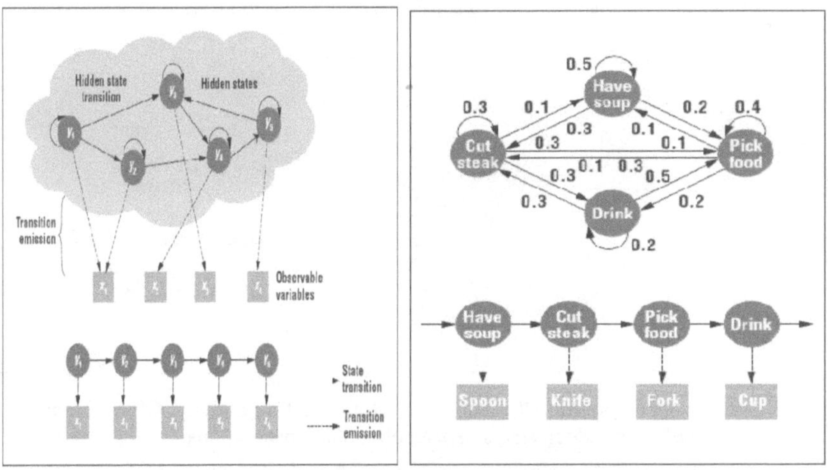

Fig. 6.8 The graphical representation of a hidden Markov model (HMM). By observing sequences of sensor data (observable variables), the HMM discovers hidden state transitions. HMM for an eating activity, Based on the objects used (spoon, knife, fork, or cup, which are the observable variables), we can infer the HMM states and their transitions.

6.2.1.2 The Conditional Random Field (CRF)

Mostly actions show non-deterministic temperament enabling independence for a few steps of the activity to be carried out in any order. In reality most actions are synchronized and interconnected. The depending unique field shows greater conformity to HMM interacting with practical needs. The stability of invisible varying "y" on noticed varying "x" makes CRF as discriminative and generative probabilistic design.

The HMM and CRF are utilized to find invisible condition conversion from statement series. Instead of discovering a combined possibility submission p(x, y) with HMM design, CRF tries to find the only depending possibility p(y|x). CRF allows for very subjective, reliant relationship amongst statement series adding to with versatility. One of the huge difference between HMM and CRF design is the reduction of independence assumptions where invisible condition assumptions depends on precedent and potential findings. The CRF is created as an undirected acyclic graph that flexibly shows link amongst statement varying and invisible condition. An example of a conditional random field for an eating activity is shown in figure 6.9.

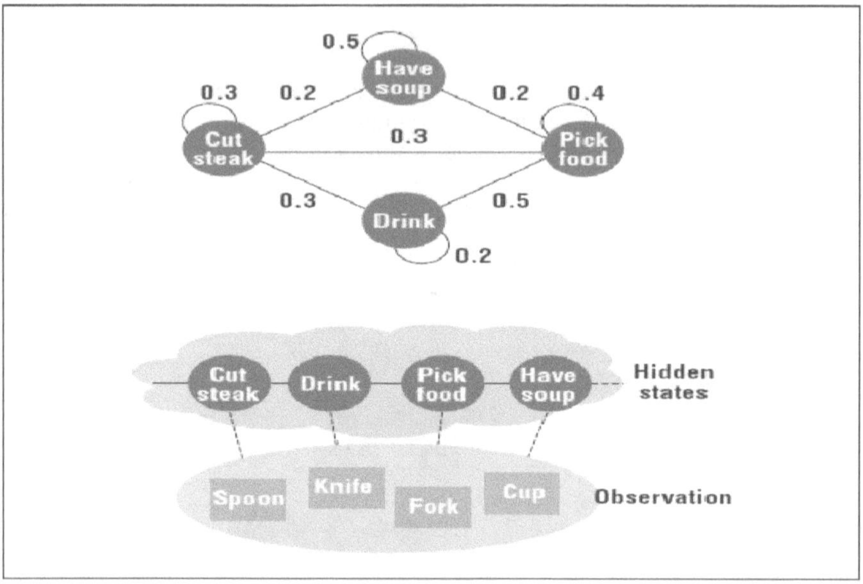

Fig. 6.9 Example of a conditional random field for an eating activity. Observations aren't randomly generated, and hidden states depend on global observations

A CRF uses a potential function instead of a joint probability function. Suppose there are hidden variables Y= (y1, y2,…., yt−1, yt) and observation variables X=(x1,x2, …, xk). The two probabilities, transition and observation, of the HMM are replaced respectively by a transition feature function r (y_{t-1}, y_t), and a state feature functions (y_t ,X). Both feature functions return 1 in the simplest case if there's a correlation between its variables. The potential function p(Y | X) is computed by the equation:

$$p(Y|X)=\frac{1}{Z(X)}\exp\left(\sum_i \lambda_i \sum_{t=1}^{n} f_i(y_{t-1},y_t,X,t)\right),$$

where

$$f_i(y_{t-1},y_t,X,t)=r(y_{t-1},y_t) \text{ or } s(y_t,X),$$

λi is a weight of correlation that represents the actual potential. In the eating example, the λi values shown in Figure .. are estimated from training data. Z(X) is a normalization factor to convert a potential value to a probability value between 0 and 1.

6.2.1.3 The Skip Chain Conditional Random Field (SCCRF)

The operation of the method is straightforward and books towards Linear-Chain CRF. The Linear-Chain CRF is more certified than HMM because of its slackening freedom supposition. Both spread the restrictions of being suitable for activities showing successive temperament. The forming of blend activities with synchronized and interconnected sub-activities requires complicated prospective functions to sum up long- range dependency, where CRF is branded as Miss Cycle CRF or SCCRF. It's a straight line Cycle having higher space between two factors and can easily explain SCCRF prospective operate as product of six several Straight line stores. These prospective products are high-priced from the computation perspective to higher event of Miss Cycle between activities.

6.2.1.4 Emerging Patterns (EP)

In the method of emerging pattern every action shows a current design attributing a function vector for personal actions, showing key modifications amongst two segments of information. Example: a information set containing numerous cases has personal set of factors with its related principles. Within the current function, few hold up a class more than others. Example: the function vector {location cooking area, object@burner} is a cooking action growing design and {object@cleanser, object@plate, place kitchen} is a growing design for cleaning of home action. These features can be located by processing support and rate of growth for personal features.

The quantity of machine studying designs used for action identification significantly differences similar to sort of actions and indicator information determined and used. The successful action identification is obtained by using Naïve Bayes classifiers. The classifier recognizes the actions managing with the highest possibility to the range of noticed indicator principles which is expected as conditionally separate. Few scientists have involved decision trees to study understandable, reasonable details of the actions.

All the synchronized and interconnected actions are determined by Emerging Routine methods. Apart from Emerging Styles, every method needs structured studying which consequently requires training information for true action identification like greater limitations in carrying out designed labeling. The modifications in indicator environment must be corresponded in the person designs.

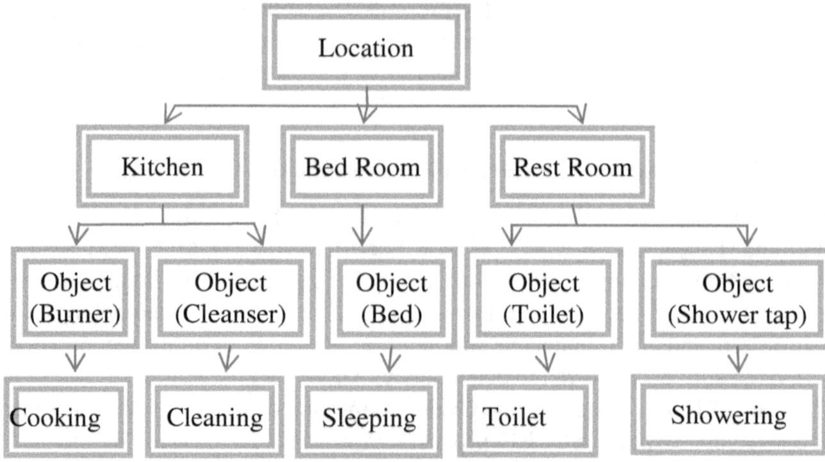

A fragment of the decision tree, modeled based on the smart home sensor data

Leaf nodes of the decision tree correspond to the activities of an inhabitant in a smart home.

A set of predictive rules, modeling the data set is:

Location= Kitchen and Object (Burner) => Cooking

Location=Kitchen and Object (Cleanser) =>Cleaning

Location=Bedroom and Object (Bed) =>Sleeping……

Selected descriptive rules, describing individual patterns in the data are:

Location=Object(Burner)=>Cooking,

Location=Object=(Cleanser)=>Cleaning,

Location=Object(Bed)=>Sleeping……

From a semantic point of view, emerging patterns are association rules with an itemset in rule antecedent, and a fixed consequent: ItemSet→D1, for given data set D1 being compared to another data set D2[6,7].

The measure of quality of emerging patterns is the growth rate (the ratio of the two supports).

It determines, for example, that a pattern with a 10% support in one data set and 1% in the other is better than a pattern with support 70% in one data set and 10% in the other (as 10/1 >70/10).From the association rule perspective, GrowthRate(ItemSet,D1,D2)= Confidence(ItemSet→D1)/1−confidence(ItemSet→D1).

Consider a data set D that consists of several instances, each having a set of attributes and corresponding values. Among the available attributes, some support a class more than others. For instance, a feature vector {location@kitchen, object@burner} is an EP of a cooking activity and {object@ cleanser, object@plate, location@kitchen} is an EP of a cleaning a dining table activity To find these attributes, support and GrowthRate are computed for every attribute X[6,7]:

$$Supp(X) = \frac{\text{the number of instances containing } X \text{ in } D}{\text{the number of instances in } D}.$$

Given two different classes of data sets $D1$ and $D2$, the growth rate of an item set X from $D1$ to $D2$ is defined as

$$GrowthRate(X) \begin{cases} 0 & \text{if } \mathrm{Supp}_1(X) - 0 \text{ and } \mathrm{Supp}_2(X) - 0 \\ \infty & \text{if } \mathrm{Supp}_1(X) = 0 \text{ and } \mathrm{Supp}_2(X) > 0 \\ \frac{\mathrm{Supp}_2(X)}{\mathrm{Supp}_1(X)} & \text{otherwise} \end{cases}.$$

EPs are item sets with large growth rates from D1 to D2. These EPs are mined from sequential sensor data and applied to recognize interleaved and concurrent activities [6,7].

6.2.2 Finding Patterns in Sensor Data

A further understanding of raw sensor information can be obtained by implementing knowledge discovering methods to automatically recognize precedents of interest in the information. Example: "Association Analysis" finds ongoing co-occurrences amongst reports in the raw sensor information. The co-occurrences duplicate connection helping in comprising connection as sensible. Sensor information includes temporary elements. Time-ordered sensor information can be used to investigate a repeated series that possibly matches with styles of importance.

A window of user-defined sizing is changed by certain criteria by using sensor information to generate, selection periods or development of interests. Subject's episode can be evaluated through the rate of repeat, periodicity and duration or by utilizing lowest possible information duration concept.

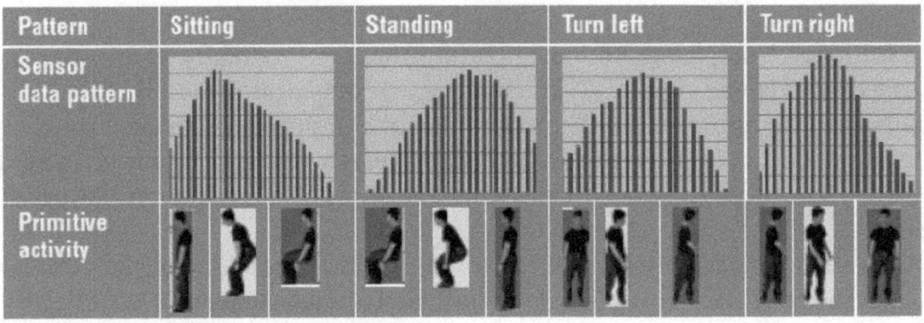

Fig. 6.10 Extraction of motion patterns from sensor data. Each video key frame is composed of three primitive poses [7].

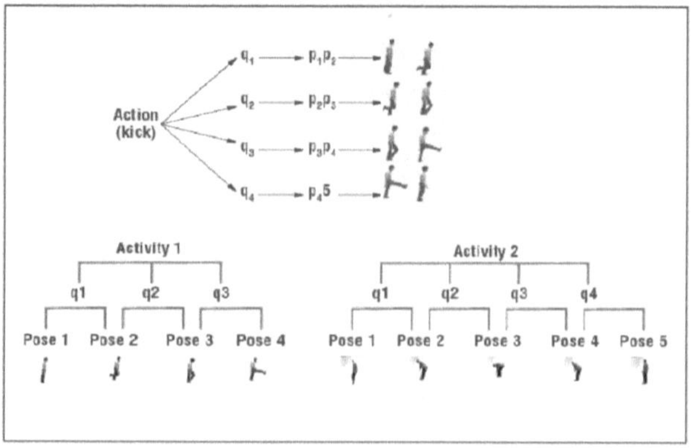

Fig. 6.11 Composite activities are a combination of multiple poses according to a sequence rule [7]

6.2.3 *Classifying Sensor Data*

Primary of data-analysis is to data out indicator information points to defined category brands that classify the information. Example: depiction of task is carried out to identify an individual living at smart home through the research of time-ordered information obtained from motion receptors. Different sorts of classification methods is related to indicator information which includes decision shrub, Bayesian classifier, sensory networks, instance-based students, support vector machine and regression methods.

Categorization methods differs with regards to the nature of information managing like category brands having unique or true principles, information made up of overlooked principles or discrepancies, scale of training information accessibility and the reflection of educational principles. Health care must be taken to avoid the over suitable of learning criteria with the information leading to non-generalization of perceptive concept to the latest information.

6.2.4 *Detecting Trends*

The additional assessment of indicator information eventually elements can be performed to decide upon styles that point out principles along specified component of the information. Various styles like improve, decrease, cyclic styles and continuous conditions are shown. The design research method uses date autocorrelation plots showing link among time-shifted principles in an occasion series.

The continuous design under patience limit includes continuous autocorrelation principles with in the shown information that is being analyzed. Cyclic design matches with greater cyclic principles predicting greater upwards mountains in an autocorrelation information. In case of increasing and reducing styles, greater level of autocorrelation is observed among nearby and near-adjacent findings. At lag one there is greater connection which progressively reduces with improve in the lag. The path of change can be established by identifying the difference in the connecting factors. This function of study to identify styles in the of health position centered on indicator information, collected at smart home is very effectively utilized by the scientists.

There is a unique type of design called "anomaly detection" also determined as "outlier detection". Flaws are information factors which do not match with basic model. The abnormality identification can be made by performing various mathematical assessments like Grubbs test which decides the odd Z score between the information factors.

6.2.5 Characterizing Sensor Data

Encounters with higher quantity of indicator information require recognition of styles in the information that helps in distinguishing the characteristics of information and few major designs. The information depiction can be accomplished through different sorts of methods available in higher numbers. One criteria categories the information into unique sets which segregates the information factors into individual group like individuals in an identical group are more as well than the information factors of different group.

K-means is one of the most common clustering methods with initial companies of k categories being chosen at unique and the left over information factors are being assigned to "nearest" group as the design of distance features must suit with information. The final allowance of information factors to specific group causes the recalculation and reassigning of group companies. The process carries on until group balance is achieved.

6.2.6 Annotation Methods

A view of action identification which has been considerably under research is the method acquainted to describe example selection that the researchers will implement to information the action design. Several researchers have released the results of tests where the participants are needed to observe every action. While in some other cases, the researchers said the participants with particular actions ought to be performed, so the actual action brands were identified previously the indicator details was even collected. In one case, the detective examined the raw indicator details to describe it with an comparative action brand. For all circumstances, none of these faces are realistic. Hand labeling from the indicator details is a time intensive and hence may not be best strategy either.

Usually, find out actions that normally need often in personal ads home environment after that create designs to recognize actions what they happen. Based on result, there is no information description of action details is required. By realizing while happen these actions, sensible atmosphere and PCPs (primary care providers) methods will evaluate how consistently, frequently and completely personal ads are performing their regular performances.

Discover the action styles which are regular for personal instead of using chosen actions. Also, to focus on inter topic variation issue. Also implement action building to set the styles in to action details, in which group centroids represent the actions which can be determined and monitored. Therefore, create a marketed edition of a disguised Markov design to represent actions and their changes hence to recognize these actions whenever they happen in Smart Environment.

References

[1] Arabnia, H.R., Fang, W.-C., Lee, C., Zhang, Y.: Context-Aware Middleware and Intelligent Agents for Smart Environments. IEEE Intelligent Systems 25(2), 10–11 (2010), doi:10.1109/MIS.2010.47

[2] Jae Hyuk, S., Boreom, L., Kwang, S.P.: Detection of Abnormal Living Patterns for Elderly Living Alone Using Support Vector Data Description. IEEE Transactions on Information Technology in Biomedicine 15(3), 438–448 (2011)

[3] Tibor, B., Mark, H., Michel, C.A.K., Jan, T.: An ambient agent model for monitoring and analysing dynamics of complex human behavior. Journal of Ambient Intelligence and Smart Environments 3(4), 283–303 (2011)

[4] McKeever, S., Ye, J., Coyle, L., Bleakley, C., Dobson, S.: Activity recognition using temporal evidence theory. Journal of Ambient Intelligence and Smart Environments 2(3) (2010)

[5] Wood, A., Stankovic, J., Virone, G., Selavo, L., He, Z., Cao, Q., Doan, T., Wu, Y., Fang, L., Stoleru, R.: Context-aware wireless sensor networks for assisted living and residential monitoring. IEEE Network 22(4), 26–33 (2008)

[6] Rashidi, P., Cook, D.J., Holder, L.B., Schmitter Edgecombe, M.: Discovering Activities to Recognize and Track in a Smart Environment. IEEE Transactions on Knowledge and Data Engineering 23(4), 527–539 (2011)

[7] Cook, D.J.: Learning Setting-Generalized Activity Models for Smart Spaces. IEEE Intelligent Systems 27(1), 32–38 (2012), doi:10.1109/MIS.2010.112

[8] Hu, D.H., Yang, Q.: Cigar: Concurrent and interleaving goal and activity recognition. In: AAAI 2008: Proceedings of the Twenty-Third AAAI Conference on Artificial Intelligence, pp. 1363–1368 (2008)

[9] Zhongna, Z., Wenqing, D., Eggert, J., Giger, J.T., Keller, J., Rantz, M., He, Z.: A real-time system for in-home activity monitoring of elders. In: Proceedings of the Annual International Conference of IEEE Engineering in Medicine and Biology Society, EMBC 2009, September 3-6, pp. 6115–6118 (2009)

[10] Suryadevara, N.K., Mukhopadhyay, S.C.: Wireless Sensor Network based Home Monitoring System for Wellness Determination of Elderly. IEEE Sensors Journal 12(6), 1965–1972 (2012)

[11] Gaddam, A., Mukhopadhyay, S.C., Gupta, G.S.: Elder Care Based on Cognitive Sensor Network. IEEE Sensors Journal 11(3), 574–581 (2011)

[12] Ma, T., Kim, Y., et al.: Context-Aware Implementation based on CBR for smart home. IEEE Trans. Wireless and Mobile Computing, Networking and Communications, 112–115 (2005)

[13] Acampora, G., Loia, V., et al.: Ambient intelligence framework for context aware addaptive applications. In: Proc. of the 7th International Workshop on Computer Architecture for Machine Perception (2005)

[14] Liminh, C., Chris, D.N., Wang, H.: A knowledge-Driven Approach to Activity Recognition in Smart Homes. IEEE Trans. Knowledge and Data Engineering 24(6), 961–974 (2012)

[15] Sanchez, D., Tentori, M.: Activity Recognition for the Smart Hospital. IEEE Intelligent Systems 23(2), 50–57 (2008)

[16] Bao, L., Intille, S.: Activity Recognition from Userannotated Acceleration Data. In: Proc. Int'l Conf. Pervasive Computing, pp. 1–17 (2004)

[17] Brdiczka, O., Crowley, J.L., Reignier, P.: Learning Situation Models in a Smart Home. IEEE Trans. Systems, Man and Cybernetics—Part B: Cybernetics 39(1), 56–63 (2009)

[18] Hoey, J., Poupart, P.: Solving POMDPs with Continuous or Large Discrete Observation Spaces. In: Proc. Int'l Joint Conf. Artificial Intelligence, pp. 1332–1338 (2005)

[19] Brand, M., Kettnaker, V.: Discovery and Segmentation of Activities in Video. IEEE Trans. Pattern Analysis and Machine Intelligence 22(8), 844–851 (2000)

[20] Robertson, N., Reid, I.: A General Method for Human Activity Recognition in Video. Computer Vision and Image Understanding 104(2), 232–248 (2006)

[21] Ward, J.A., Lukowicz, P., Troster, G., Starner, T.E.: Activity Recognition of Assembly Tasks Using Body-Worn Micro hones and Accelerometers. IEEE Trans. Pattern Analysis and Machine Intelligence 28(10), 1553–1567 (2006)

[22] van Kasteren, T.L.M., Noulas, A.K., Englebienne, G., Krose, B.J.A.: Accurate Activity Recognition in a Home Setting. In: Proc. 10th Int'l Conf. Ubiquitous Computing, UbiComp (September 2008)

[23] Hu, D.H., Yang, Q.: CIGAR: Concurrent and Interleaving Goal and Activity Recognition. In: Proc. Am. Assoc. for Artificial Intelligence (AAAI) Conf. (2008)

[24] Helal, S., Mann, W., El-Zabadani, H., King, J., Kaddoura, Y., Jansen, E.: The Gator Tech Smart House: A Programmable Pervasive Space. Computer 38(3), 50–60 (2005)

[25] Cook, D.J., Rashidi, P.: Keeping the Resident in the Loop: Adapting the Smart Home to the User. IEEE Trans. Systems, Man, and Cybernetics J., Part A 39(5), 949–959 (2009)

[26] Bao, L., Intille, S.S.: Activity Recognition from User-Annotated Acceleration Data. In: Ferscha, A., Mattern, F. (eds.) PERVASIVE 2004. LNCS, vol. 3001, pp. 1–17. Springer, Heidelberg (2004)

[27] Brdiczka, O., Crowley, J.L., Reignier, P.: Learning Situation Models in a Smart Home. IEEE Trans. Systems, Man and Cybernetics—Part B: Cybernetics 39(1), 56–63 (2009)

[28] Suryadevara, N.K., Gaddam, A., Rayudu, R.K., Mukhopadhyay, S.C.: Wireless Sensors Network Based Safe Home to Care Elderly People: Behaviour Detection. Elsevier Sensors and Actuators: A Physical, `http://dx.doi.org/10.1016/j.sna.2012.03.020`; Procedia Engineering 25, 96–99 (2012)

[29] Yu, Z., Nakamura, Y.: Smart Meeting Systems: A Survey of State-of-the-Art and Open Issues. ACM Computing Surveys 42(2), Article 8 (2010)

[30] James, A.B.: Activities of Daily Living and Instrumental Activities of Daily Living. In: Crepeau, E.B., Cohn, E.S., Schell, B.B. (eds.) Willard and Spackman's Occupational Therapy, pp. 538–578. Lippincott, Williams and Wilkins (2008)

[31] Hamid, R., Maddi, S., Johnson, A., Bobick, A., Essa, I., Isbell, C.: A Novel Sequence Representation for Unsupervised Analysis of Human Activities. Artificial Intelligence 173(14), 1221–1244 (2009)

Chapter 7
Description of a Few Projects

Introduction

Based on the knowledge of the previous chapters a few projects have been developed. A short description of the projects is given in this chapter. The following projects are described:

1. Physiological parameters monitoring system,
2. Emotion recognition system,
3. Smart power management system.

7.1 WSN Based Physiological Parameters Monitoring System

A ZigBee based wearable non-invasive device has been developed to monitor physiological parameters (such as body-temperature, heart-rate, detection of fall) of a human subject. The system consists of an electronic device which is worn on the wrist and finger, by the person to be monitored continuously. The system can be used by an elderly or the person at risk or even by a normal person for the monitoring of physiological parameters. Using several sensors to measure different vital signs, the person can be wirelessly monitored within his own home. A heart-rate sensor has been developed to monitor the heart rate continuously. An accelerometer has been used to detect falls. The system may be extended to determine the stressed condition of the person and may be used to send an alarm signal to a receiver unit that is connected to a computer. This sets off an alarm which can go to a care-giver, allowing help to be provided to the person immediately. Since no vision sensors (camera or infra-red) are used, the system is non-invasive, respects privacy and has attracted wide acceptance especially among the elderly. The system can be used in combination with the bed sensor (part of the home monitoring system) to monitor the person during the night. The complete system will help to monitor the person during day and night and will be suitable for monitoring elderly person living alone at home.

A paper has been published on this work as is provided in reference [1]. The readers may get a quick overview of the work from reading the paper.

S.C. Mukhopadhyay: *Intelligent Sensing, Instrumentation and Measurements*, SSMI 5, pp. 141–175.
DOI: 10.1007/978-3-642-37027-4_7

Figure 7.1 shows the block diagram of the developed physiological parameters measuring system. The physiological parameters measuring sensors are designed and the signals from the sensors are interfaced with a microcontroller for processing. The UART of microcontroller sends the sensor data to the ZigBee module through serial communication. The data is then sent wirelessly to the ZigBee coordinator which then stores the data in a computer for future use.

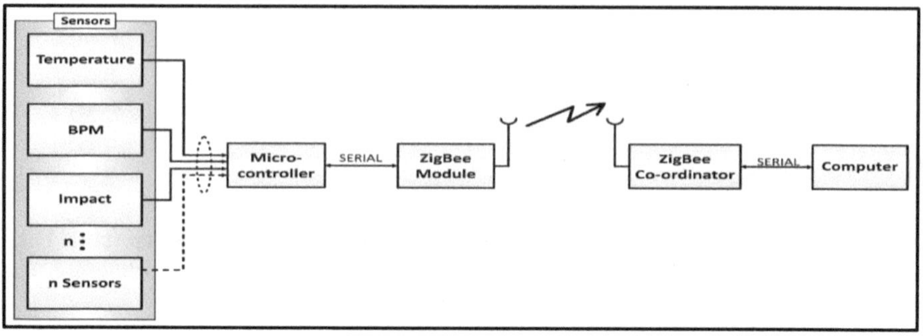

Fig. 7.1 Block diagram of the developed physiological parameters monitoring system

The constraints of the design are:

i. It should be compact and non-intrusive.
ii. It must be convenient enough to be able to wear everyday and should not be a burden for the user. The monitoring of the patient's vital signs should have to be non-invasive in order for them to be able to wear it all the time and not feel uncomfortable.
iii. It must be robust and have a user friendly interface for the caregiver to communicate with it properly.

The final prototype is shown in figure 7.2, and on a user in figure 7.3. The sensors PCB are fitted into the enclosure, which is mounted on a wrist band. The enclosure has holes on one side through which inputs of heart rate sensor, temperature sensor enter into the enclosure. The header connector inside the enclosure is connected to a ribbon cable which then takes all the sensor signals to the C8051F020 microcontroller unit. The PCB is powered by 3.3 V power rail of the microcontroller board, connected through the ribbon cable.

A 9 V battery was connected to the microcontroller board. This provided power for the sensor unit and ZigBee module through 3.3 V power rail on microcontroller analog and digital ports.

The ZigBee module connected to microcontroller consumes 40 mA during transmission. However XBee modules have the option of going in sleep mode while not transmitting.

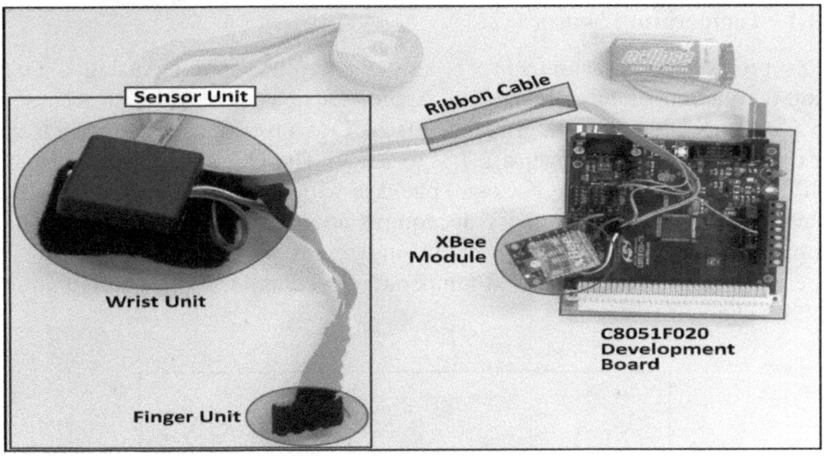

Fig. 7.2 Fabricated physiological parameters monitoring unit

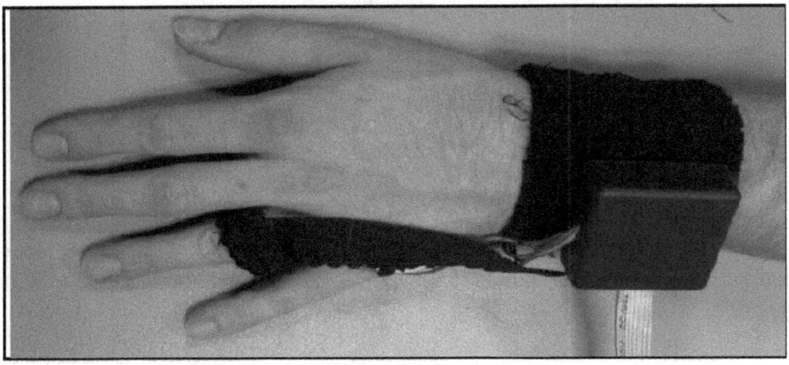

Fig. 7.3 Final Prototype unit on a user

7.1.1 Measurement of Human Body Temperature

Temperature is the most often-measured environmental quantity. Body temperature is not fixed, but is responsive to cyclical changes. The body temperature for an average active human increase during the day reaching a maximum point by evening and then lowers to a minimum point in early morning. The body temperature is an estimate of the average temperature of the core portions of the body as reflected by the temperature of the blood in the major vessels. Traditionally, body temperature has been measured by contact thermometers in the oral, rectal or auxiliary sites. These sites are choices of convenience rather than correctness, because they often do not represent internal (core) body temperatures with the required accuracy.

7.1.1.1 Temperature Sensor

In this project the measurement of body-temperature is done using a DS600 analog-output temperature sensor from Analog Devices. It provides an accuracy of ±0.5°C over its wide temperature range of -20°C to +100°C. This accuracy is valid over its entire operating range of 2.7 V to 5.5 V. The DS600 is available in 8-pin µSOP (Micro Small Outline Package) package with an exposed pad on the bottom of chip for quick thermal response. It requires no external components and has an exposed pad which helps in making maximum skin area in contact with temperature sensor to get the best temperature reading. The pin configuration of DS600 is shown in figure 7.4.

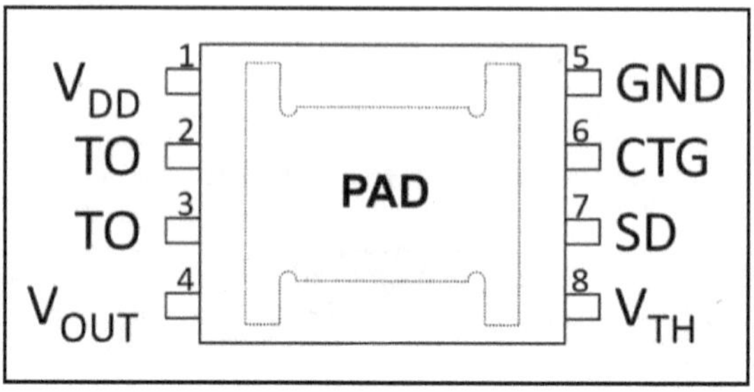

Fig. 7.4 Top view of DS600 Pin Configuration

A typical application circuit is shown in figure 7.5. The sensor is powered with 3.3 V from the microcontroller board. Pin no. 4 gives output voltage, which is proportional to the die temperature in degrees centigrade and Pin no. 8 is for ground connection.

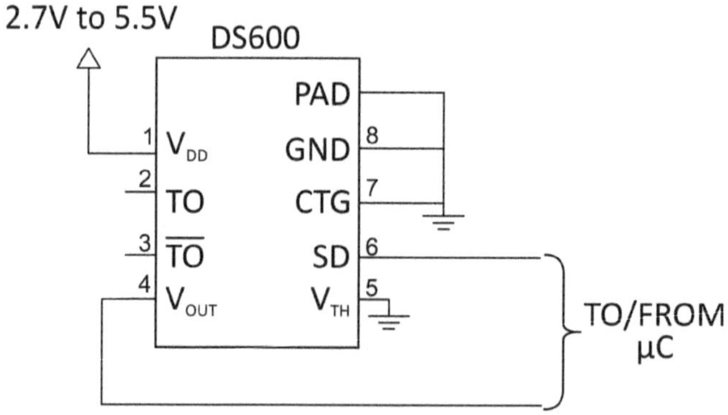

Fig. 7.5 A typical circuit of using DS600

DS600 measures its own temperature and provides these measurements in the form of analog output voltage. It is mentioned in the data sheet that output voltage characteristics is factory calibrated for a typical output gain of (ΔV/ΔT) of +6.45mV/°C and a DC offset (VOS) of 509mV. Over the required range the sensor has an accuracy of ±0.5°C. To calibrate the sensor, the DC signal was measured and the corresponding ADC value on microcontroller is noted at varied ambient temperature. The figure 7.6 shows the transfer function of the laboratory experimental readings using this sensor of human skin temperature against the output voltage.

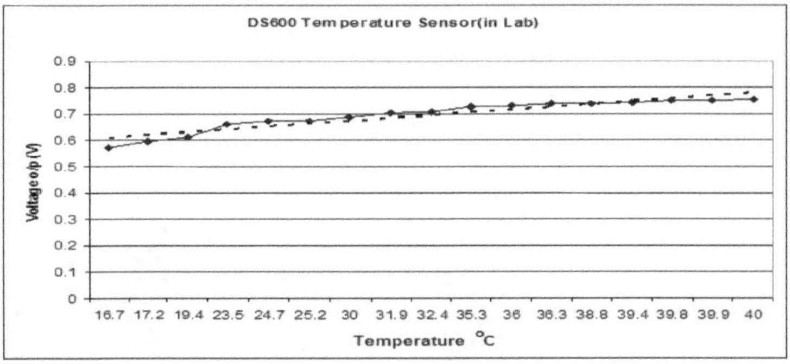

Fig. 7.6 Voltage output characteristics of DS600

This is compared with original DS600 Transfer function as shown in figure 7.7. The typical output voltage transfer characteristic equation is:

$$V_{out} = (\text{Device Temperature } (°C) * (\Delta V/\Delta T) + V_{os}).$$

The sensor is mounted within the wrist strap, positioned in such a way that it is in direct contact with the skin, allowing it to measure the external temperature of the body. To get quick thermal response from the sensor, the exposed pad is placed on a PCB with precision cut PCB hole. The IC is then mounted and soldered on the PCB, such that the bottom part of the chip is exposed to get maximum exposure to the skin surface under temperature measurement. The PCB layout of the IC is shown in Figure 7.8.

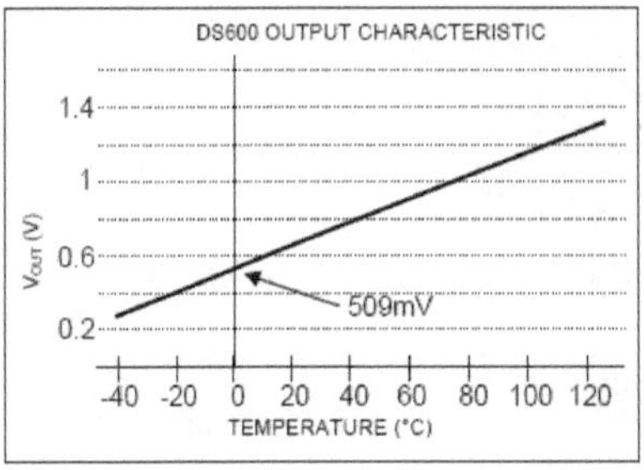

Fig. 7.7 Transfer function of DS600

Fig. 7.8 DS600 IC on PCB

For collecting experimental data each sensor was tested separately. Though the temperature measurements were taken at several, different parts of the body, the place that was chosen for later "wearing" the monitoring unit is the wrist. This seemed to be the most comfortable and less annoying position due to the reason that people are used to wear anything at this part of the body i.e. a wristwatch or a bracelet.

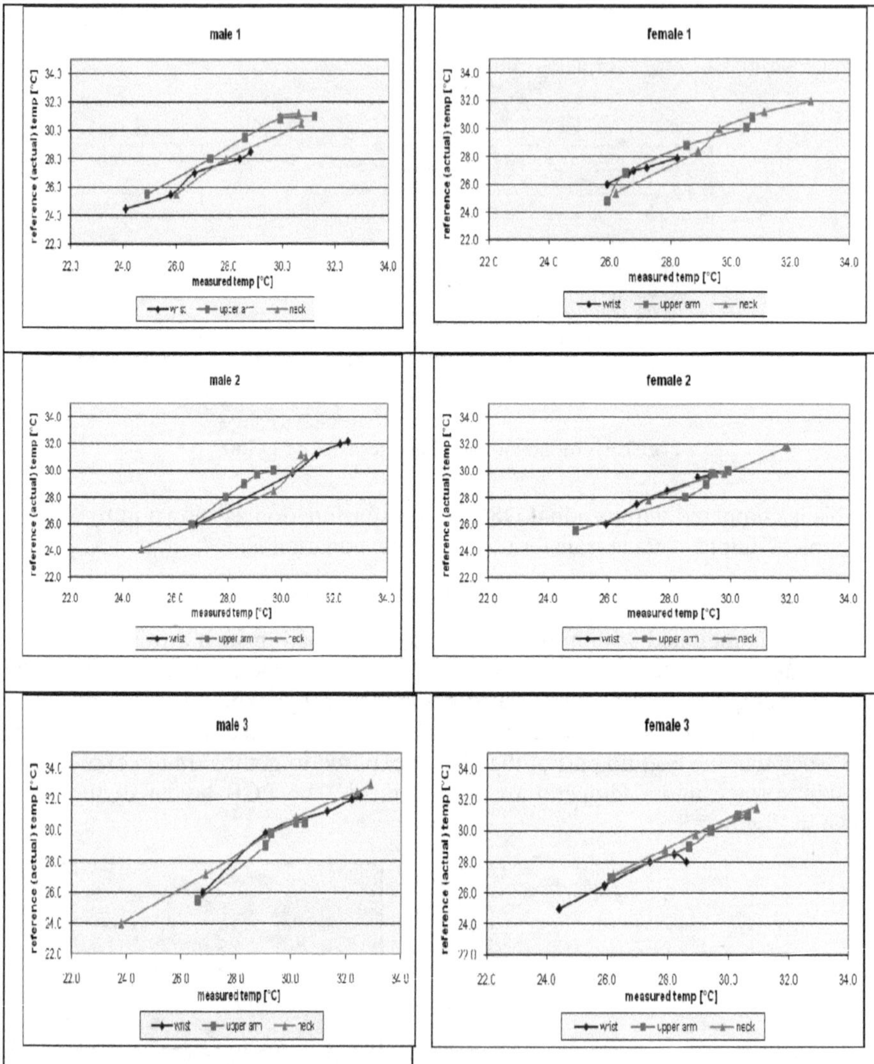

Fig. 7.9 Test results of temperature measurements on 3 male and 3 female subjects

For verifying temperature, the sensor PCB is placed on wrist, then on upper forearm and on neck. The readings are obtained at different time intervals. The temperature at three different positions (wrist, neck and upper arm) was measured for three male and three female persons. The temperatures were measured at different times with varying ambient conditions. Figure 7.9 shows the variations in temperature measured with respect to different positions.

The figure 7.10 shows the errors in the measurement of body temperatures.

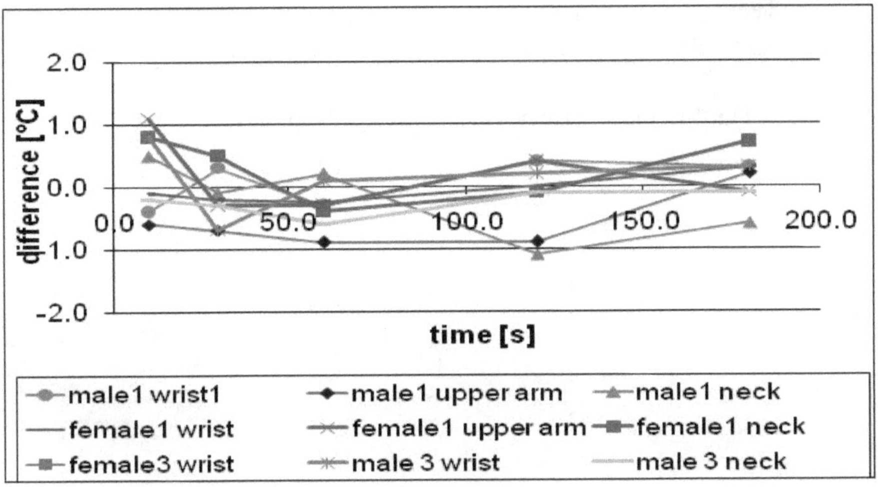

Fig. 7.10 Experimental results of errors in temperature measurements

7.1.1.2 The Heart-Rate Sensor

For the measurement of heart-rate infrared based approach is used. An infra-red light source is directed at the patient's skin, and the amount of light that is reflected is measured by a photo-diode. Infrared light is absorbed by hemoglobin but not by most other tissue present in the human body. Hence, the amount of light detected varies with the pulse of the patient (as the amount of blood increases and decreases as a function of heart pulse). This can be measured to determine the heart rate of a patient.

A custom heart rate sensor was designed to read the patient's heart beats per minute (bpm). Near-infrared spectroscopy based non-invasive approach has been used to determine the heart rate. It consists of a Gallium Arsenide (GaAs) infrared LED with a wavelength of 940nm. It transmits the signal through the tissue and is received by an infra red sensor-Phototransistor (SDP8406). The time between the impulses are measured and counted so that frequency is obtained. The block diagram is shown in figure 7.11 and the circuit diagram is shown in figure 7.12.

A silicon phototransistor (SDP8406), moulded into a flat side-facing package and a GaAs Infrared Emitting Diode were used in the sensor. For testing purposes first 5mm GaAs infrared LED was used, which later on was replaced by 1.9mm axial LED, shown in figure 7.13. The axial LED was mounted on a PCB for better hold on to the fabric, used for holding and positioning senor around finger.

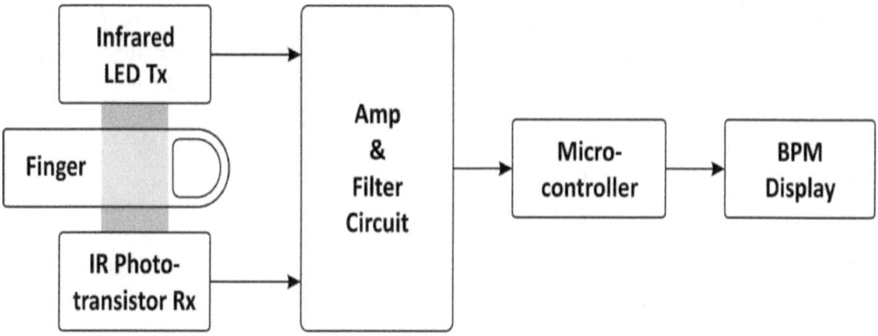

Fig. 7.11 Block diagram representation of the Heart Rate sensor

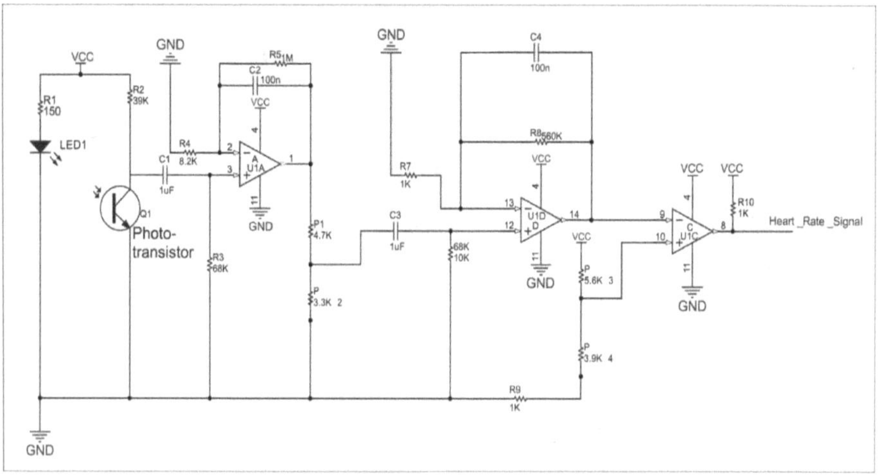

Fig. 7.12 Circuit diagram of the Heart Rate sensor

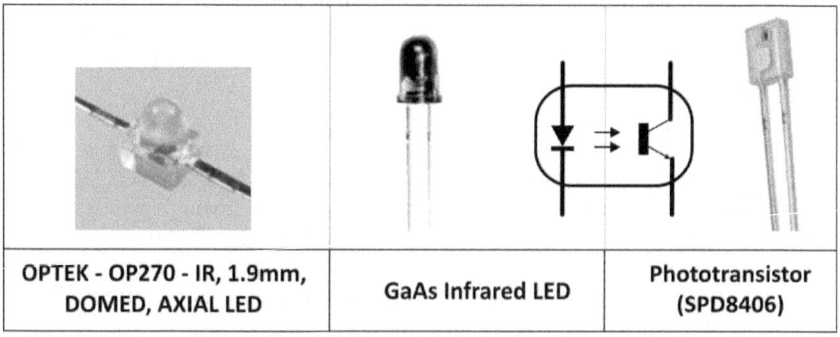

Fig. 7.13 Types of LED's, phototransistor used in the system

Table 7.1 Comparison of experimental Results of heart rate measurement

Subject	Heart Rate (BPM)		Reference: Pulse watch (BPM)		Error (BPM)	
	Male	Female	Male	Female	Male	Female
1	58	55	57	54	+1	+1
2	78	59	77	57	+1	+2
3	54	58	55	58	-1	0
4	72	68	71	71	+1	-3
5	68	65	67	64	+1	+1
6	75	58	75	59	0	-1
7	69	64	69	64	0	0
8	85	87	86	86	-1	+1
9	79	77	81	79	-2	-2
10	70	61	69	62	+1	-1
11	75	92	77	93	+2	-1
12	65	58	64	58	+1	0

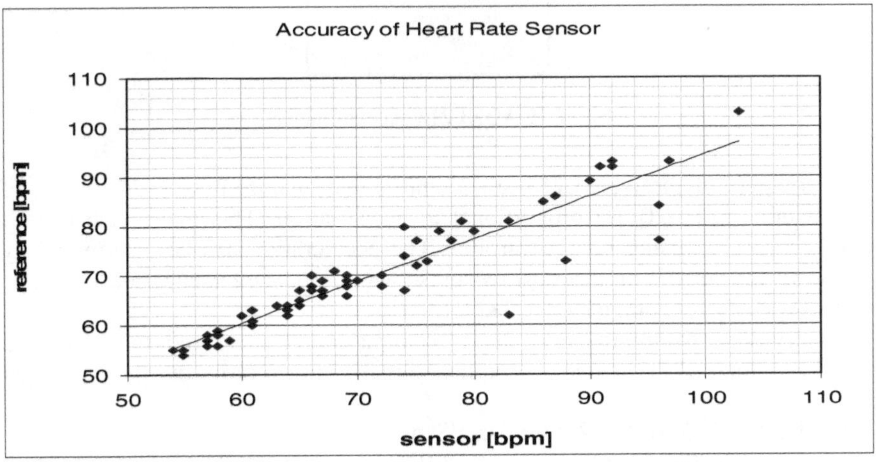

Fig. 7.14 Accuracy of measurement of the Heart Rate sensor

A sport watch (WR30M) with a similar sensor was used as reference for comparison. Heart rate measurement data were collected simultaneously. The comparative results in tabular form are shown in table 7.1 and in the graphical form in figure 7.14.

7.1.1.3 Impact Sensor

Impact or fall detection sensors are ideal for sensing whether or not a person has fallen over, indicating serious injuries. Injuries sustained from falls can be fatal resulting in serious physical and physiological consequences. Injuries may result in broken bones, superficial cuts to the skin and may include damage to underlying bone tissue. In most countries worldwide, falls are the leading cause of death or injury-related hospitalisation among people of 65 years and older in society.

A fall detecting sensor in this project has been implemented by using an accelerometer. The ADXL213 as shown in figure 7.15 from Analog Devices was used, capable of measuring dynamic and static changes in acceleration in the horizontal and vertical axis. This sensor operates by using small plates suspended by springs, which deflect when subjected to acceleration/gravity. The ADXL213 was chosen as it was more readily available and provided static force measurement.

The ADXL213 is a complete dual axis acceleration measurement system on a single monolithic IC, containing a poly-silicon surface-micro-machined sensor and signal conditioning circuitry to implement open-loop acceleration measurement architecture. The ADXL213 is capable of measuring both positive and negative accelerations with a dynamic range of ±1.2 g.

Fig. 7.15 ADXL213 accelerometer

The outputs of the ADXL213 are digital signals whose duty cycles (ratio of pulse width to period) are proportional to acceleration. This duty cycle can be directly measured using microcontroller. The acceleration can be determined by measuring the length of the positive pulse width (t_1) and the period (t_2). The nominal transfer function of the ADXL213 is:

Sensitivity = Minimum magnitude of input signal required to produce a specified output signal having a specified signal-to-noise ratio, or other specified criteria.

Acceleration = ((t1/t2) – Zero g Bias)/Sensitivity

In the case of the ADXL213 , Zero g Bias = 50% nominal and Sensitivity = 30%/g nominal.

Figure 7.16 shows the necessary circuit diagram to use the ADXL213.

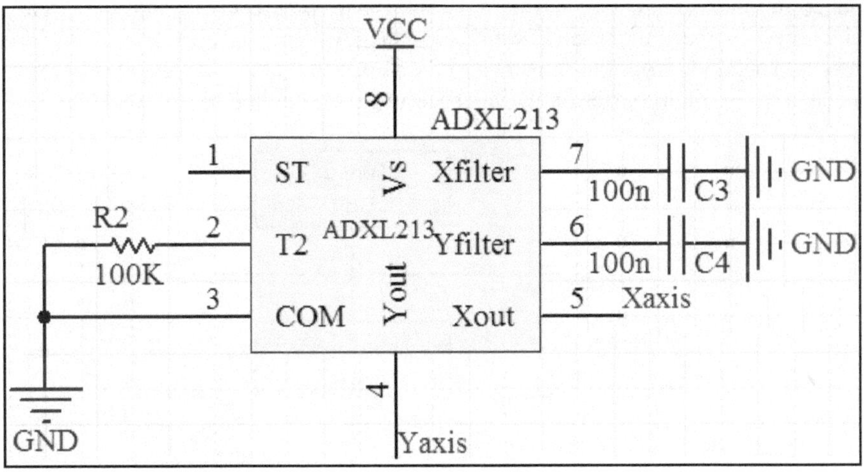

Fig. 7.16 Circuit Diagram of motion detection sensor

The output of the accelerometer was tested with different conditions. Figure 7.17 displays the various movements of the user in respect to duty cycle (%).

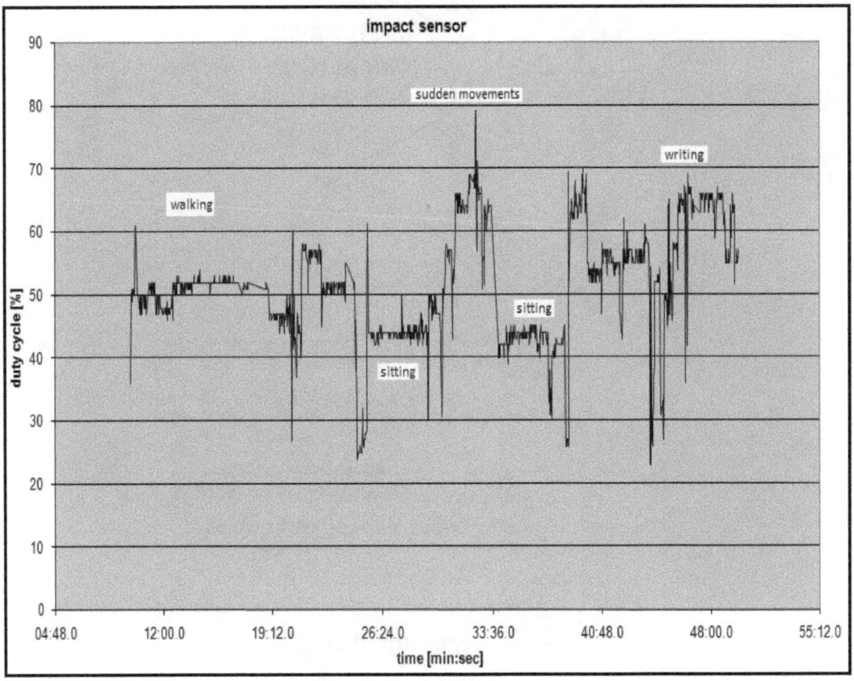

Fig. 7.17 The measured duty cycle for various positions of user

7.1.1.4 Communication between Sensor Unit and Micro-controller

The communication between the wrist unit and the receiver unit is wireless, transmitted in the unlicensed 2.4 GHz frequency band. Information is gathered every two seconds from the sensors and then encoded into a packet. This packet is then sent to the radio buffer on the microcontroller, and then transmitted.

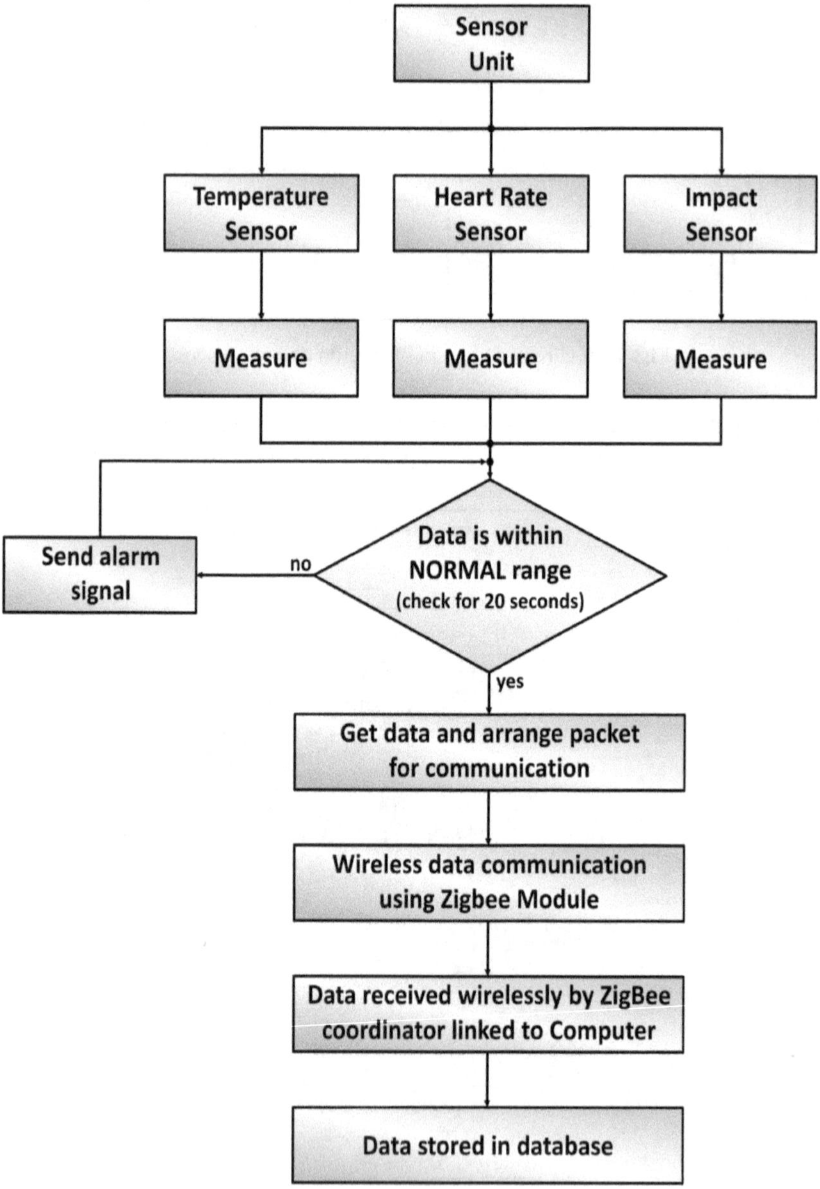

Fig. 7.18 Communication algorithm

The flowchart in Figure 7.18 displays how the communication takes place between sensor unit, microcontroller and ZigBee.

7.1.1.5 Software and Algorithms

This section provides some information about the design algorithms used in the project.

Wrist Unit Microcontroller: The software for the C8051F020 has been written in an evaluation version of SDCC complier, which allows up to 64k bytes of object code.

The developed system software is based on three sensors which are designed to measure physiological parameters of body, the transmission of data through ZigBee.

The code is designed on an interrupt driven basis. All interrupts are triggered using the timers in auto-reload mode, triggering an interrupt every time the timer reaches its maximum value. At certain pre-defined intervals algorithms are run to gather information from the sensors, perform some processing and assemble the information into a packet.

7.1.1.6 Temperature Sensor Algorithm

Analog signal from the temperature sensor is connected to ADC0 of C8051F020 microcontroller. The converted ADC values are taken in the form of arrays. 50 sample values are taken and measured, threshold filtering is also introduced. The data values above and below this threshold value are discarded to remove spurious inputs. Average of the 'good' data is calculated using the equation

$$Device\ Temperature\ (°C) = (ADCout - 509)*10000/645$$

Where ADCout (mV) = (Result/16) *2430/4095, Result is the 12-bit digital output of the ADC0.

7.1.1.7 Impact Sensor Algorithm

The figure 7.19 illustrates how the digital output of the impact sensor is used to determine the duty ratio by the microcontroller.

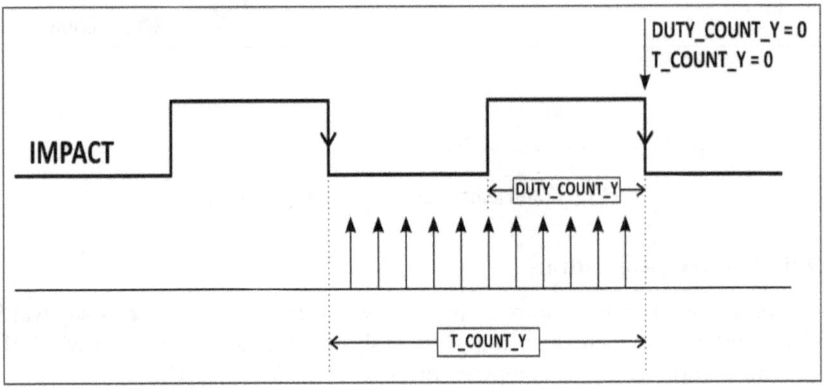

Fig. 7.19 Interfacing of impact sensor with the microcontroller

Sensor signal (PWM) is fed to pin #3 of port0 (external hardware interrupt INT1) of the microcontroller. Timer0 generates "software ticks" every 10μs. Timer0 ISR is used to increment the T_COUNT_Y and also increment DUTY_COUNT_Y only if the sensor signal is high. The width of the pulse (DUTY_COUNT_Y) and time period (T_COUNT_Y) are calculated.

And the duty cycle is given by:

Duty Cycle = (DUTY_COUNT_Y *100)/T_COUNT_Y or this can be written as

Dutycycle = dutycount * 100/Tcount value.

7.1.1.8 Heart Rate Sensor Algorithm

In a similar way the signal of the heart rate sensor is interfaced to the micro-controller as is shown in figure 7.20.

The sensor signal (pulses) is fed to pin #2 of port0, using external hardware interrupt INT0 of the microcontroller. Due to which Timer0 generates "software ticks" every 10 μseconds. Timer 0 Interrupt Service Routine (ISR) keeps incrementing the value of BPM_T_count value. INT0 ISR counts the number of software ticks between two negative edges of the incoming sensor signal. Frequency is counted using the equation:

Frequency = 10000000/BPM_T_count_value and to get the heart rate per minute, following equation is used:

BPM = Frequency*6/10.

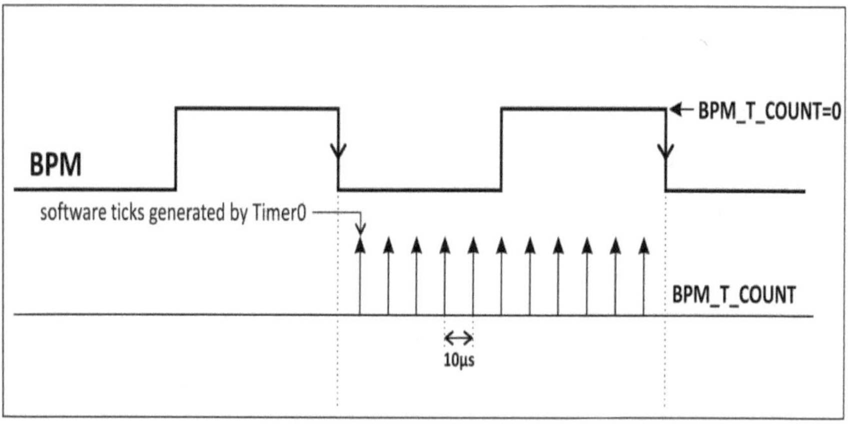

Fig. 7.20 Heart rate (BPM) sensor interface

7.1.1.9 Receiver Algorithms

The receiver unit is much simpler than the wrist unit. It is in a constant waiting state for information, forwarding wirelessly to the serial port. The ZigBee coordinator is connected to computer through a USB cable. This is illustrated in figure 7.21.

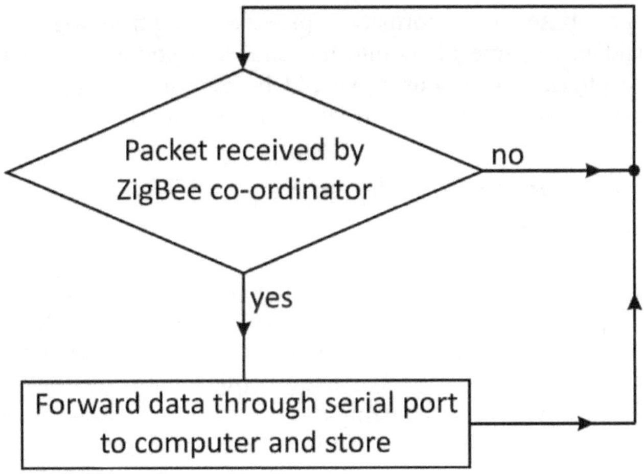

Fig. 7.21 Receiver unit process

The program is a user interface, allowing a report on the current status of the individual. Once the user has connected to the receiver unit, data is automatically updated on the screen. BPM, temperature and impact (in both axes) is given on the display.

7.2 Intelligent Sensing System for Emotion Recognition

The development of the physiological parameters monitoring system has been re-arranged for recognition of human emotions. Emotions play a vital role in people's everyday life. It is a mental state that does not arise through free-will however it is often accompanied by physiological changes. Emotions play a central role in decision making, problem solving, communicating, negotiating, and adapting to unpredictable environments. It consists of external physical expression as well as internal feelings, thoughts and internal processes which the person experiencing the emotion may not be aware of. Emotions can be categorized into various types. The six basic emotions widely accepted and used in literature are happiness, sadness, fear, surprise, anger, and disgust. Other emotions that are not commonly used include amusement, contempt, contentment, embarrassment, excitement, guilt, pride, achievement, relief, satisfaction, sensory pleasure, and shame. Monitoring emotions is important as it contains information that can help in improving human wellbeing. Emotion recognition has also become an important subject when it comes to human-machine interaction.

7.2.1 Aim of the Emotion RecognitionSystem

The main aim of this project is to develop a low-cost intelligent real-time monitoring system capable of evaluating four basic emotions i.e. happy, neutral, sad and angry, based on information provided by physiological sensors. The system should be capable of monitoring data in a comfortable and unobtrusive manner. The physiological sensors should be non-invasive and power efficient. The system should be capable of wireless communication.

7.2.2 Development of Intelligent Sensing System for Emotion Recognition

A human emotion recognition system based physiological parameters has been developed. The physiological parameters provide the inner feelings rather than only physical changes. Since the physiological changes are controlled by the autonomous nervous system which is largely affected by emotions, so monitoring these changes will help in recognizing the basic four emotions. Heart rate sensor, skin conductance sensor and skin temperature sensor are used for continuous real time data monitoring and consequently recognition of emotion.

The heart rate sensor used in this project is based on the principle of photoplethysmography (PPG). PPG is an optical measurement technique used for detecting blood volume changes in the micro vascular bed of tissue. This method uses a light source (LED) which illuminates the skin and a light detector (photo diode) that detects the changes in optical properties due to change in blood volume. One way is to have the light detector and light source on the same side of the skin, this is called reflectance PPG. Reflective PPG works by illuminating a bed of capillaries, typically with light in the infrared range due to its low absorption by tissue, before measuring how the back-scattered light is modulated by the blood. The other way is to have light detector and light source placed across the skin; this is called transmittance PPG. Transmittance PPG usually occurs at the extremities of body including fingertips, toes and earlobes due to large density of capillaries. In this project transmission PPG is used as it is more reliable, resistant to movements and external noise and also provides a comfortable monitoring of heart rate. In this project a 940nm Optek OP180 LED was used as the light source while for light detection, SDP8406 phototransistor by Honeywell was used. Both the LED and phototransistor have low cost, low power consumption and compact size. Electronic circuit using LM324 (quad op-amp) was designed for amplification and A/D conversion of the signal from the phototransistor. The digital signal is fed into the analogue port of the microcontroller. The microcontroller used in this project is a C8051F020 by SiLabs which processes the analogue and digital signals from the sensor.

The skin conductance response sensor based on a simple voltage divider rule where the human body acts as a medium between two electrodes. These electrodes come in contact with the fingers where the change in resistance or conductance is observed. In this project the conductance is observed with equivalent electrical signal values ranging from 100mV-1500mV which is fed into the analogue port of the microcontroller for processing. According to the literature these values are dependent upon the level of arousal which in turn is related to human emotions.

The skin temperature is measured by a DS600, Maxim - Dallas semiconductor as explained before.

Once all the sensors were designed and tested, the next step was to integrate them together for data collection. In order to achieve this, the sensors and electrodes were placed on the surface of a box shape casing where they were easily accessible by the left human hand for comfortable data monitoring. The heart rate data is collected from the index finger, the skin conductance response electrodes are in contact with the middle and ring finger while the skin temperature data is collected from the palm under the flexor digiti mini brevis muscle. Figure 7.22 shows the system casing design for data monitoring and human emotion analysis.

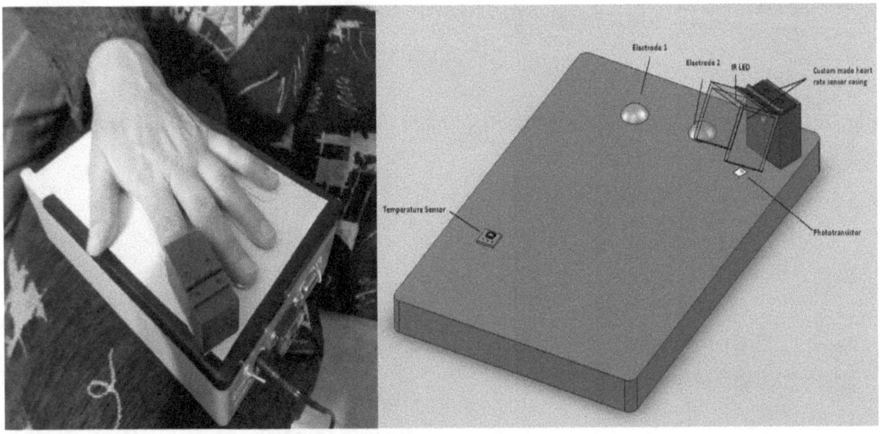

Fig. 7.22 Hardware part of the emotion recognition system (left) and 3D model (right)

Once the sensors were integrated and tested, the next step was to transmit the physiological sensor data over a medium for the display and storage on a computer. The communication between computer and the sensors was brought by using ZigBee wireless technology. Two XBee modules, coordinator and router, were configured for transmission and reception of data respectively. Once the XBee modules were configured and data was transmitted wirelessly, a graphical user interface (GUI) was designed for data display, storage and analyses. For this reason a GUI using Visual Studio C# was designed, capable of real-time, data monitoring, storing, analyzing and emotion evaluation. The GUI was programmed to calculate first and second derivatives of each signal. Along with the derivatives, the absolute value for the first derivative of the heart rate signal was obtained. These values helped in providing more information about each signal and were displayed as a string on a non graphical window. Before the GUI was programmed for emotion recognition, training data was obtained for many individuals (both male and female) with ages ranging from 19 to 72 years. The data was initially analyzed visually for observing patterns and trends. Figures 7.23, 7.24, 7.25 and 7.26 show the heart rate, first derivative of heart rate, skin conductance and skin temperature of individuals observed in four emotional states.

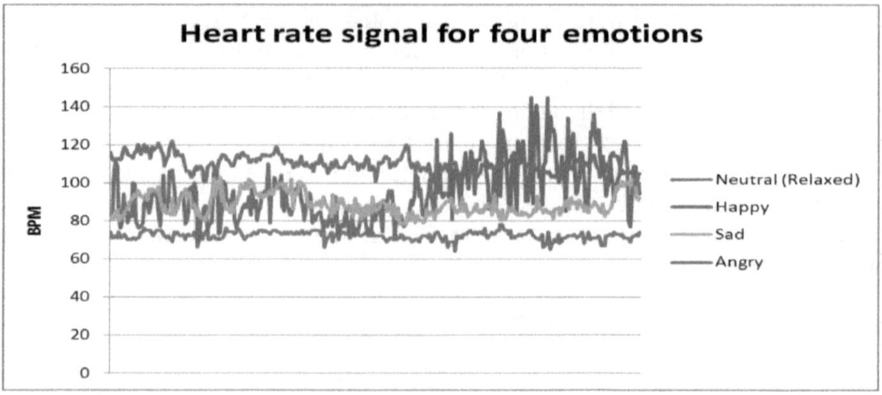

Fig. 7.23 Heart rate observed for subjects in happy (excited), sad, neutral and angry states

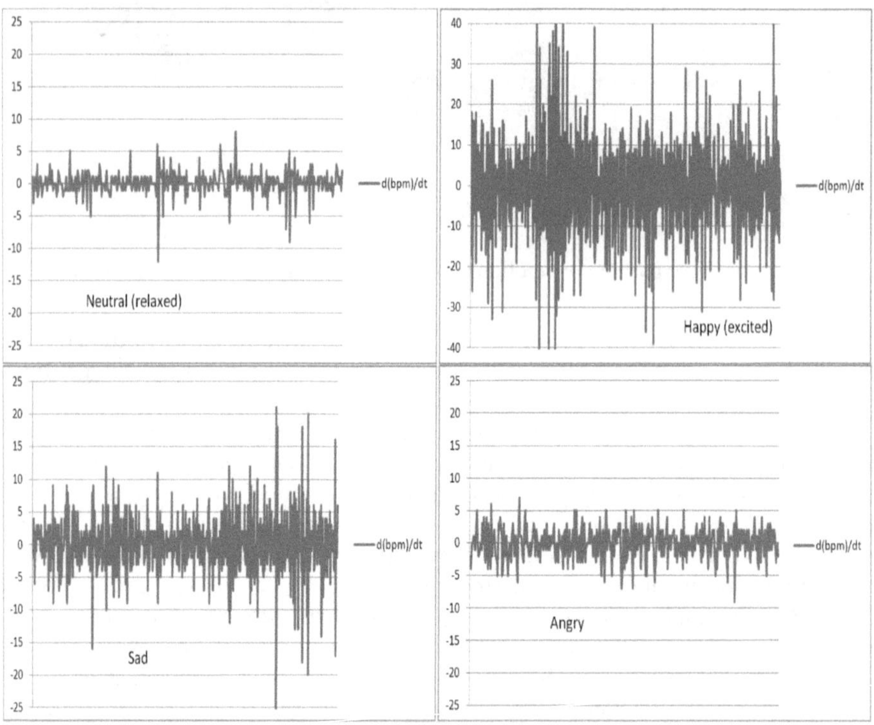

Fig. 7.24 First derivative test of heart rate signal observed for subjects in neutral (relaxed), happy (excited), sad, and angry states

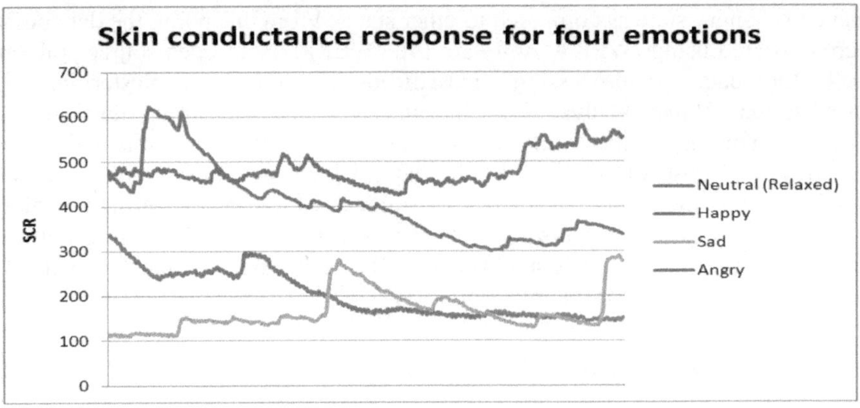

Fig. 7.25 Skin conductance response (SCR) of subjects in neutral, happy (excited), sad and angry states

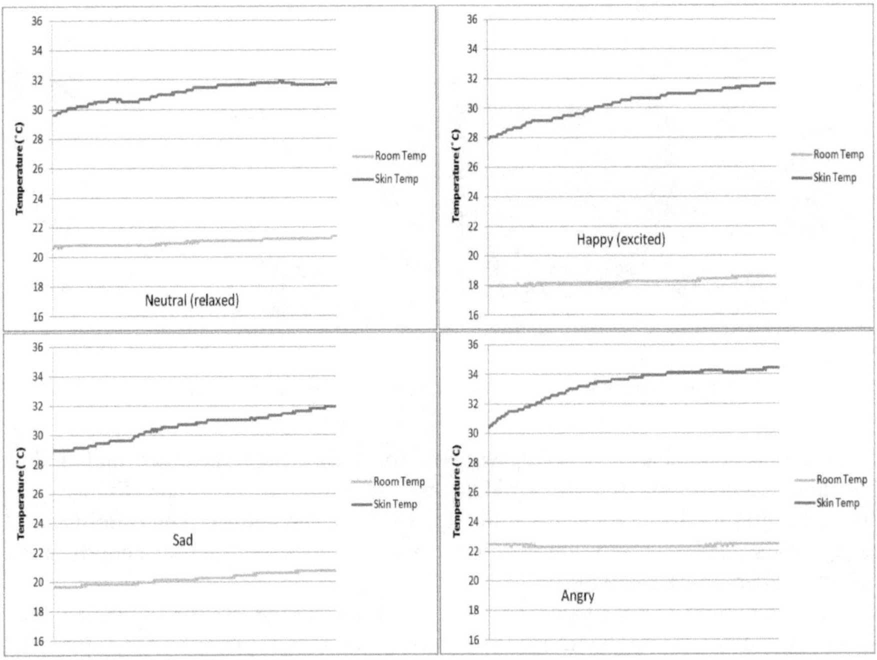

Fig. 7.26 Skin temperature observed for subjects in neutral, happy (excited), sad and angry states

The above results showed that emotions do have effect on the physiological signals which can be visualized by continuous monitoring. The heart rate variability and skin conductance response showed significant results across all four emotions. The temperature signal on the other hand showed significant change for angry state as compared to other states. From this point, the data points were clustered using 'WEKA' software [4]. 'WEKA' is an open source software used for data pre-processing, classification, regression, clustering, and visualization. K-means clustering algorithm was used to group data in four clusters (emotions). K-means clustering is a method of cluster analysis which aims to partition 'n' observations into 'k' clusters in which each observation belongs to the cluster with the nearest mean. The main idea of k-means clustering technique is to find the centres or centroids of natural clusters in the data using iterative refinement technique. The four clusters produced by WEKA matched with the visually observed data for the four emotions.

7.2.3 Experimental Results and Analysis

The XBee coordinator is connected into the computer using XBee explorer USB to collect the sensors data. Figure 7.27 shows the XBee coordinator receiving data from the XBee router/end device.

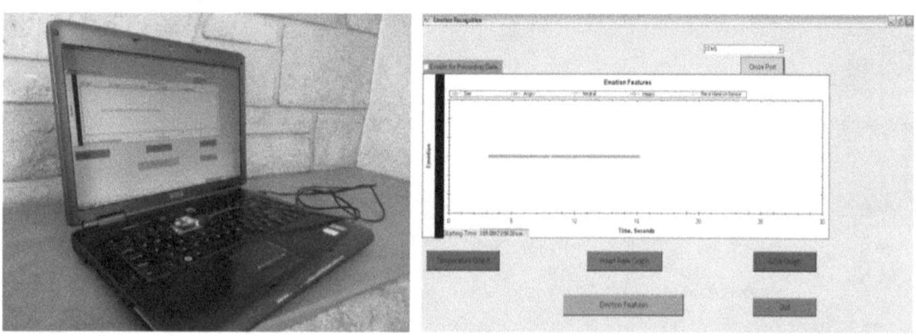

Fig. 7.27 XBee coordinator receiving data and front panel of the designed GUI

The front panel has a drop down box for COM port selection and a data recording select option for storing data as a text file for later use. Figure 7.28 shows the graphs displaying real time signals from the sensor. The amplitude of the signals is displayed on y-axis while the x-axis shows the time in seconds along with the starting time in text box.

In case the hand is not placed on the system, an orange line will appear which means that there is no hand to take data from or to place a hand for data monitoring. Figure 7.29 shows a collective representation of all four emotions displayed on the front panel of the GUI.

Data analysis is performed by following clustering exploration technique, that allows objects with similar characteristics to be grouped together in order to facilitate further processing. In this project, K-means clustering process was applied which is a popular data clustering technique.

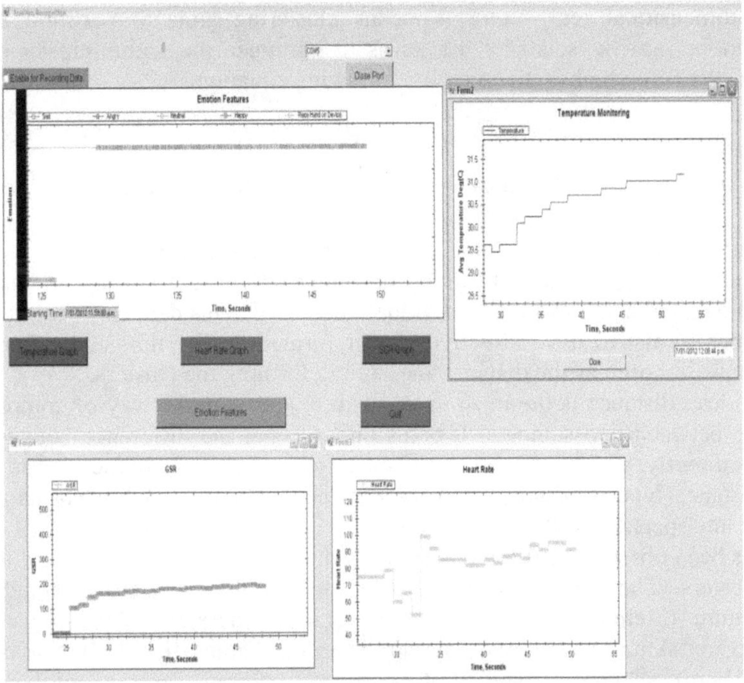

Fig. 7.28 Sensor signal graph along with the graphical real-time data

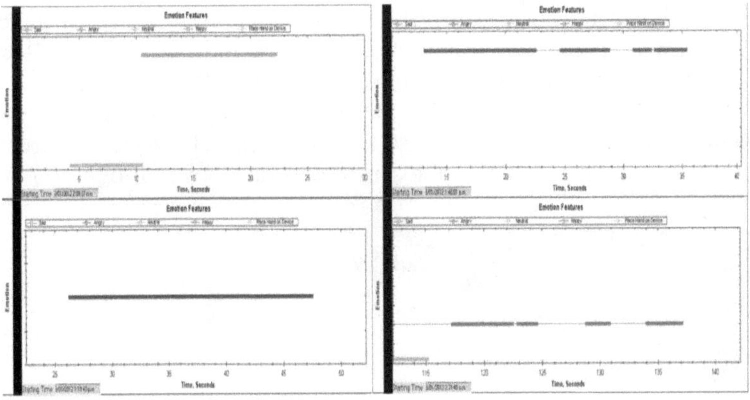

Fig. 7.29 Graphical display of all four emotions

K-means clustering is a method of cluster analysis which aims to partition 'n' observations into 'k' clusters in which each observation belongs to the cluster with the nearest mean. The main idea of k-means clustering technique is to find the centres or centroids of natural clusters in the data using iterative refinement technique. For a given a set of observations (x_1, x_2,... x_n), where each observation is a d-dimensional real vector, k-means clustering aims to partition the 'n' observations into 'k' sets ($k \leq n$), so as to minimize the within-cluster sum of squares (WCSS) which is given by the following equation.

$$J = \sum_{j=1}^{k} \sum_{i=1}^{n} \left\| x_i^{(j)} - c_j \right\|^2$$

Where $\left\| x_i^{(j)} - c_j \right\|^2$ is the distance measure between a data point ($x_i^{(j)}$)and the cluster centre c_j.

Based on the number of emotions to be recognized, it was decided to create four clusters using k-means clustering technique. In order to cluster the data, the training data collected from 40 individuals was used. This data was pre-processed for proper format to have effective cluster formation. For this study, Euclidean distance was chosen as the distance measure in forming the clusters.

Euclidean distance is the most usual, natural and intuitive way of computing a distance between two samples. It takes into account the difference between two samples directly, which is based on the magnitude of changes in the sample levels. This distance type is usually used for data sets that are suitably normalized or without any special distribution problem.

It has been observed that the basic eleven attributes (variables) are used to form four clusters. The eleven variables used for this data clustering include skin temperature (Stemp), first derivative of skin temperature (d1stemp), second derivative of skin temperature (d2stemp), heart rate (bpm), first derivative of heart rate (d1bpm), second derivative of heart rate (d2bpm), average of heart rate (avgbpm), skin conductance response (SCR), first derivative of skin conductance response (d1scr), second derivative of skin conductance response (d2scr) and finally absolute value of the first derivative of the heart rate signal. From the observed attribute details using by WEKA, we can see the effect of attributes on a particular cluster. These are skin temperature (Stemp), heart rate (Beats per Minute (BPM)), first derivative of heart rate (d1bpm), skin conductance response (SCR) and absolute value of the first derivative of heart rate. The others have minor significance as compared to these five attributes. The visual representations of the generated clusters based on the sensor data are shown in figures 7.30, 7.31 and 7.32.

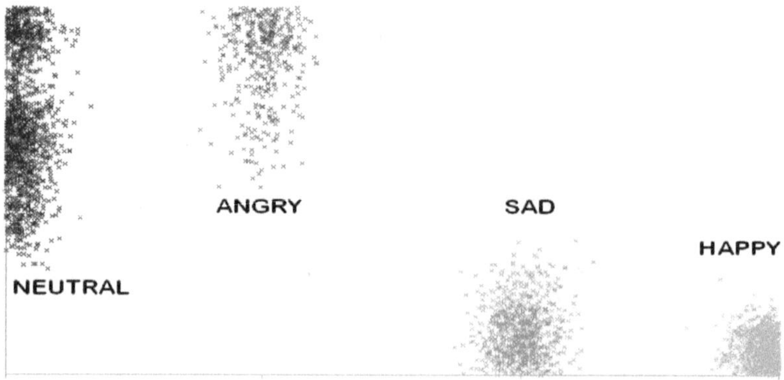

Fig. 7.30 Cluster formation for skin temperature data

Fig. 7.31 Cluster formation for Heart rate data

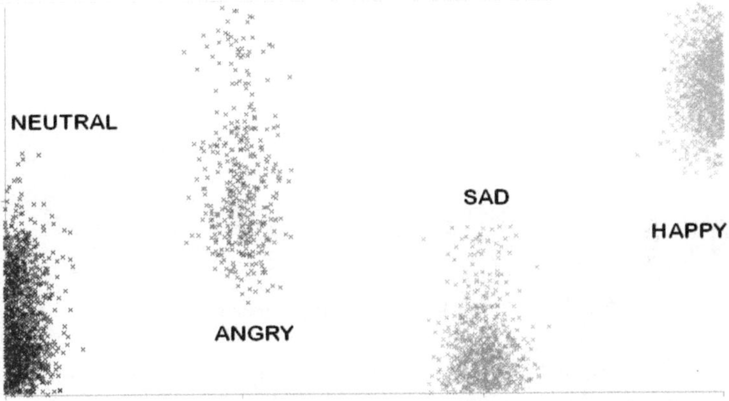

Fig. 7.32 Cluster formation for skin conductance response data

The above figures give a visual presentation of clustered data using k-means clustering technique. The emotional states are ascertained by the ground truth of the subjects participated in training and testing phase. In figure 7.30, we observed that the data points for skin temperature have the highest values for angry state and neutral states. These data points are less spread out for angry state with the centroid being towards the higher value as compared to the neutral state.

In figure 7.31, the clusters formed for heart rate data show that the angry state has data points with higher heart rate values. For happy state, the data points are widely spread compared to other states. In order to get more information out of the heart rate data, we considered the first derivative test of the heart rate data.

Similarly, the figure 7.32 depicts the clusters formed for SCR data with high values of SCR for happy and angry states while low SCR for neutral and sad states. This was true as research show that the SCR values are directly related to the arousal level of a person, therefore in angry and happy states we expect to see high SCR values while in sad and neutral state we expected a lower SCR.

Considering the five major attributes (variables) contributing to significant emotional changes values of the sensor readings, corresponding four clusters are visualized on a 2D plane as shown in figure 7.33 for effective generalization and computation. The matching clustered attribute values are specified in real time monitoring emotion recognition system for determining emotional status of the person.

The clusters formed using k-means clustering algorithm matches with the results obtained from individual emotion monitoring and data visualization. Based on the above information and conditions, the software was developed and is capable of differentiating between four different emotional states.

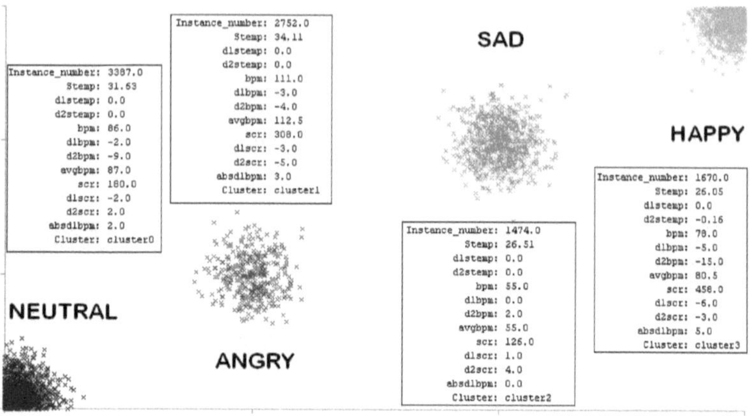

Fig. 7.33 Cluster formation using k-means algorithm

Once the emotion recognition software was completely designed and developed with the corresponding attribute cluster values, the system was tested to monitor the individual's emotional states. Forty different individuals were monitored in both natural and designed environments and achieved an overall recognition accuracy of 86.25% and their individual emotion scores are given in Table 7.2.

From the above results we can see that the neutral state achieved 100% results compared to other states. The neutral state was the reference state for other emotions and was easily recognized by the developed system. For happy state, a recognition rate of 85% was observed. During the happy state monitoring, some individuals showed emotions of excitement to a higher extent or level compared to others. Therefore, in a few instances (15%) neutral state was observed instead of the happy state. For sad emotions the lowest recognition rate of 70% was achieved. As seen from earlier data visualization and clustering, the difference between sad state and neutral state was minimal in terms of all three signals. Therefore, we did expect to get mixed results between sad and neutral state. Similarly for the angry state we achieved a recognition rate of 90%. This was because both happy and angry states are high arousal states, with high heart rate. The only visual data difference between happy and angry states was the skin temperature. Since the skin temperature is largely affected by room temperature, happy states for two individuals was observed who in reality were in an angry state (their skin temperature was low due to lower room temperature). Overall the recognition rate was found to be 86.25%.

Table 7.2 Emotion recognition rate of the developed system

Expected Emotion	Observed Emotion				
	Happy	Neutral	Sad	Angry	% Correct
Happy	34	06	0	0	85%
Neutral	0	40	0	0	100%
Sad	0	12	28	0	70%
Angry	4	0	0	36	90%

7.2.4 Observations and Discussion

In this project, a real time human emotion recognition system, based on data provided by physiological sensors has been developed. Physiological sensors were found to be the best approach to recognize emotional changes, as they provided information about changes that take place physiologically and are out of a person's control.

The heart rate sensor used in this project has been designed in a laboratory with the capability of monitoring real time heart rate and obtaining reliable data output. The skin temperature sensor used in this project measures the skin temperature from the palm area with ease and has an accuracy of ± 0.5°C. The output of the designed skin conductance response sensor has shown a direct relationship with the level of arousal. The sensors have been integrated and placed on a surface of a designed box, slightly larger than an average human hand, for collecting data from the left hand. This design enabled an easy and comfortable data monitoring system.

The four emotions showed variation in data output, a generalized statement of observation for the four emotions, with the 3 different sensors are as follows: For the neutral state, a normal heart rate value, a low rate of change heart rate, a mid-range skin conductance response and a normal to high skin temperature was observed. For the happy state, a high heart rate value, a high rate of change of heart rate, a high skin conductance response and a normal to low skin temperature was observed. For the angry state a high heart rate value, high rate of change of heart rate, a high skin conductance response and a high skin temperature was observed. For the sad state a lower than normal heart rate value, low rate of change of heart rate, a low skin conductance response and a normal to low skin temperature was observed. Along with emotion recognition and evaluation, the designed system is capable of recognizing physiological changes that arise because of various health conditions. This shows that the system has other advantages along with emotion recognition.

7.3 WSN Based Smart Power Monitoring System

A wireless sensor network based power measurement and management system for home environment has been designed and developed. The system can measure the necessary electrical parameters, such as voltage, current and power at different load points and can save the data which can be used for different purposes. Depending on the electricity tariff, the unimportant loads can be disconnected at from the system when the cost of electricity is high.

7.3.1 System Overview

Figure 7.34 shows the functional representation of the fabricated and the developed system to monitor electrical parameters and control appliances at home environment.

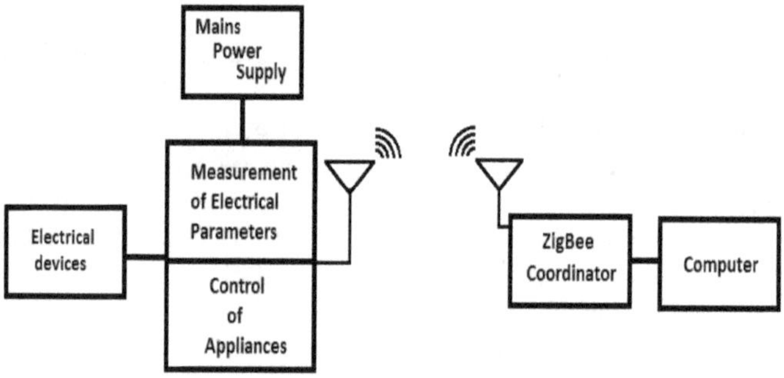

Fig. 7.34 Functional block diagram of the WSN based power management system

The voltage and currents are measured with the help of voltage and current sensors. Figure 7.35 shows the circuit diagram of voltage and current measurement. The measured signals from the sensors are integrated and connected to XBee module (end device). The sensors data then wirelessly transmitted to XBee Module (coordinator) which is connected through USB cable of the host computer, which stores the data into a database in the computer.

7.3.1.1 Voltage Measurement

The voltage transformer used in our work is the 44127 voltage step down transformer manufactured by MYRRA. Figure 7.35 shows the detailed circuit design layout for voltage measurement.

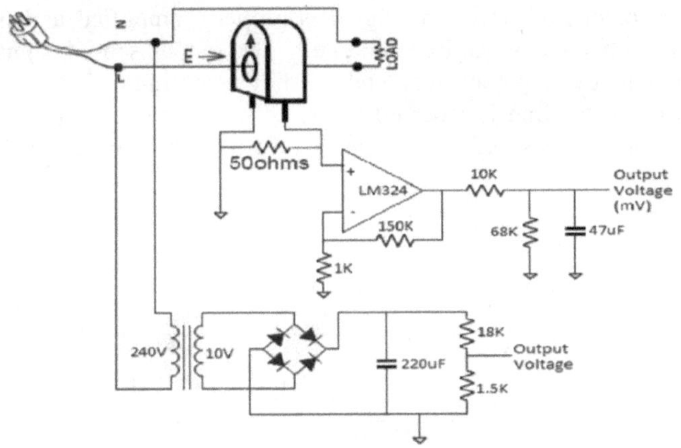

Fig. 7.35 Designed voltage and current measurement circuit

The step down voltage transformer is used to convert input supply of 230-240V to 10 VRMS AC signal which is rectified and passed through the filter capacitor to get a DC voltage. The signal is properly attenuated to bring it within the range of the ZigBee input. The output signal is then fed to analog input channel of ZigBee end device. The acquired voltage signal is directly proportional to the input supply voltage and the measured transfer characteristics are shown in figure 7.36. The scaling of the signal is obtained from the input versus output voltage graph.

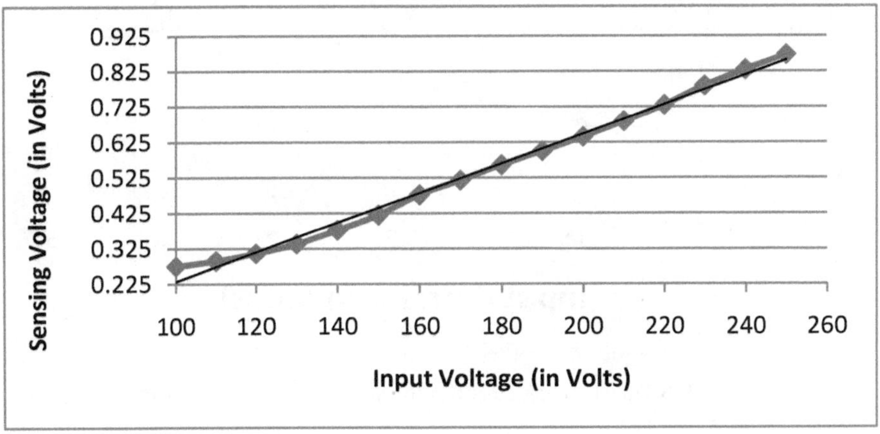

Fig. 7.36 Transfer characteristics of voltage measurement

7.3.1.2 Current Measurement

ASM010 current transformer manufactured by Talema has been used for the measurement of current. The circuit design layout for current measurement is shown in figure 7.35. In this current sensor, the voltage is measured across the burden resistor of 50 ohms. The signal is properly amplified and is then fed to analog input channel of ZigBee module. The output sensed signal is directly proportional to the input current and is shown in figure 7.37a and b for two different ranges. Scaling is required to get correct input current of the appliance. The resolution of measured signal can be improved by increasing the number of primary turns.

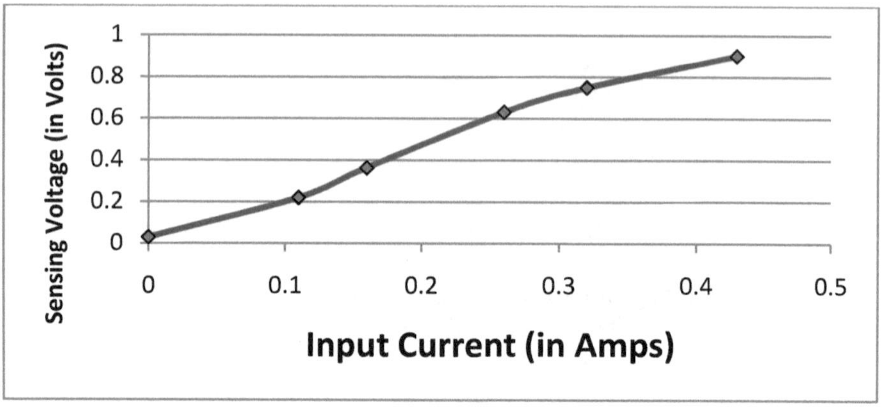

Fig. 7.37a Transfer characteristics for low current measurement

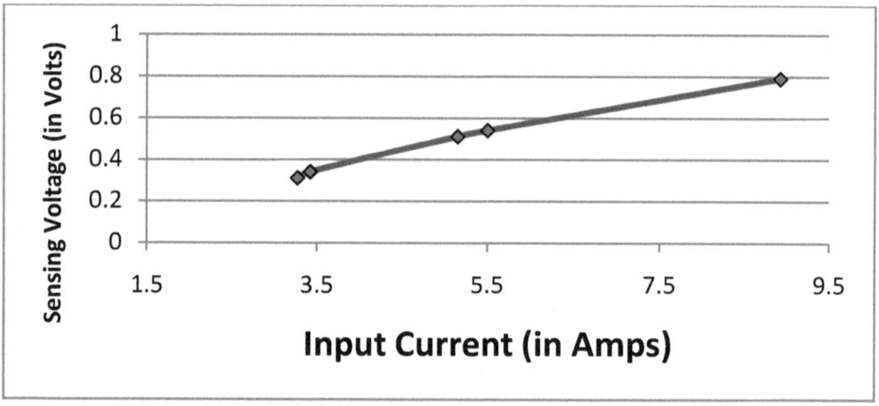

Fig. 7.37b Transfer characteristics for high current measurement

7.3.1.3 Power Measurement

The power of a single phase ac circuit is the product of line voltage and line current (rms values) and the power factor. Power Factor is the cosine of the phase angle of voltage and current waveforms as shown in the figure 7.38.

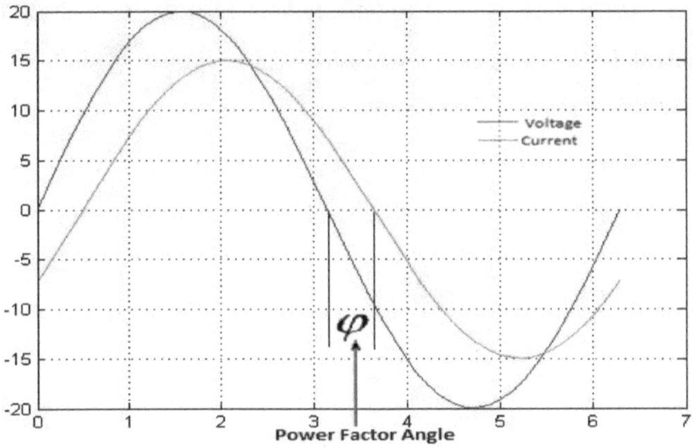

Fig. 7.38 Representation of voltage, current waveforms and power factor

The current waveforms are of proper sinusoids if the load is purely resistive, capacitive or of inductive nature. In domestic situations, most of the household appliances have some kind of electronic components which distort the current waveforms. The waveforms for a few appliances are shown in figures 7.39a, b, c, d respectively. From the graphs, it is inferred that zero-crossing determination is difficult to measure for some of the appliances and elimination of noise is not trivial. Due to the difficulty of measurement of power factor, the actual power is measured from the measurement of voltage and current. A correction factor is introduced on the product of the voltage and current measured to get the actual power.

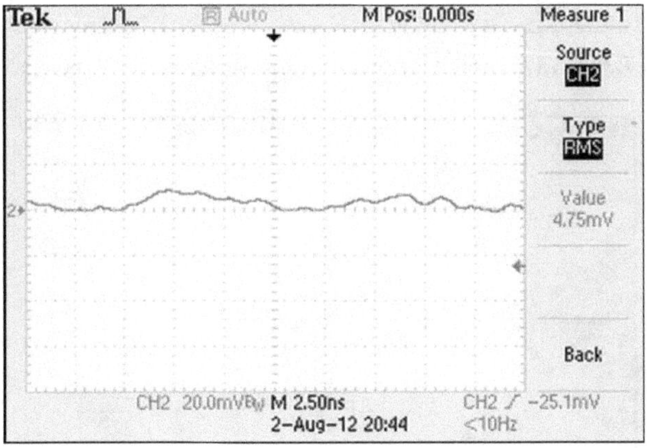

Fig. 7.39a Waveform of current transformer for a 60W electric bulb

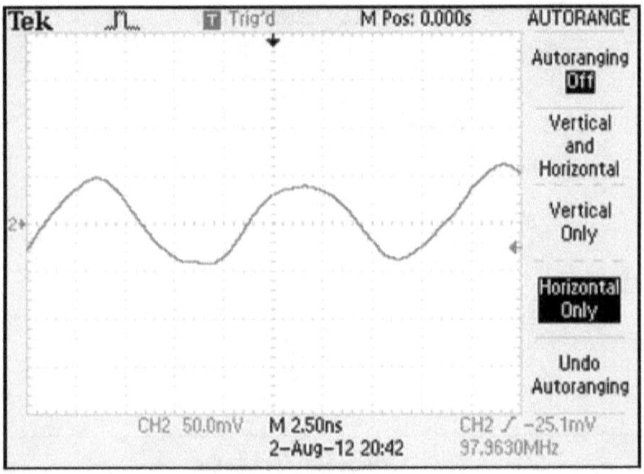

Fig. 7.39b Waveform of current transformer for a 100W electric bulb

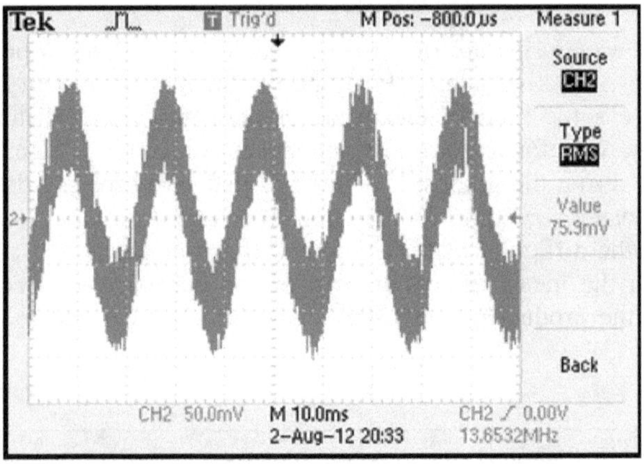

Fig. 7.39c Waveform of current transformer for a 800W room heater

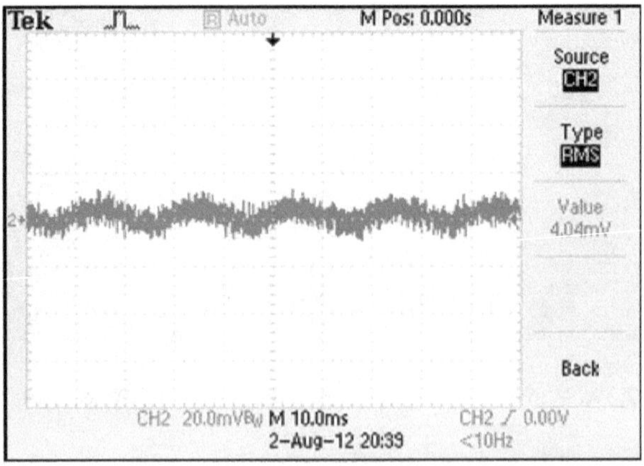

Fig. 7.39d Wave form of current transformer for an audio device

$P = v_1 * v_2 * m_1 * m_2 * C_f$;
Where, P = Calculated Power;
v_1= Output voltage from the voltage measurement circuit;
v_2= Output voltage from the current measurement circuit;
m_1=Scaling factor related to voltage measurement;
m_2= Scaling factor related to current measurement;
C_f= Correction factor;

The power is calculated in the computer after receiving voltage outputs from corresponding current and voltage sensors. Figure 7.40 shows the complete schematic diagram of the hardware part of voltage and current measurement along with the ZigBee connection.

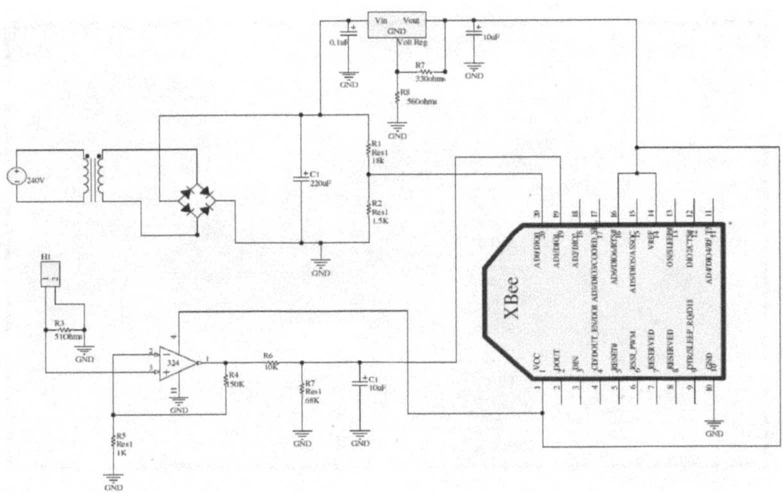

Fig. 7.40 Overall schematic of voltage and current sensing circuit integrated with ZigBee module

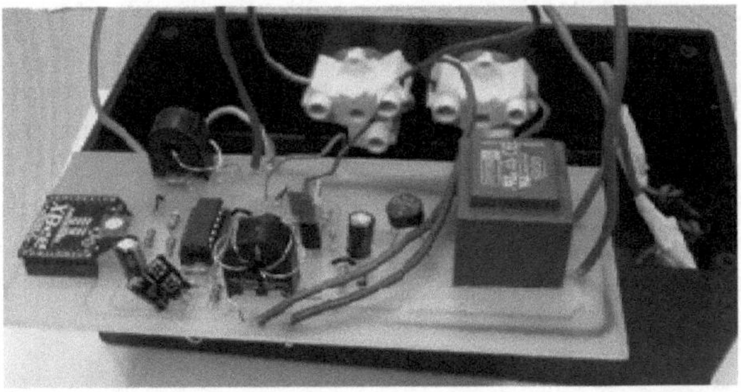

Fig. 7.41 Fabricated smart power monitoring unit

Figure 7.41 shows the fabricated system with the integrated sensing circuit and ZigBee module. The developed system includes two current transformers; one used for the measurements of loads up to 100W and the other current transformer is used for the measurements of loads from 100W to 2000W. The number of turns is increased up to five for better resolution in high loads.

7.3.1.4 Experimental Results

By monitoring consumption of power of the appliances, data is collected by a smart coordinator, which saves all data in the system for processing as well as future use. Figure 7.42 shows a graphic user interfaced (GUI) window so that appropriate action can be taken from the GUI.

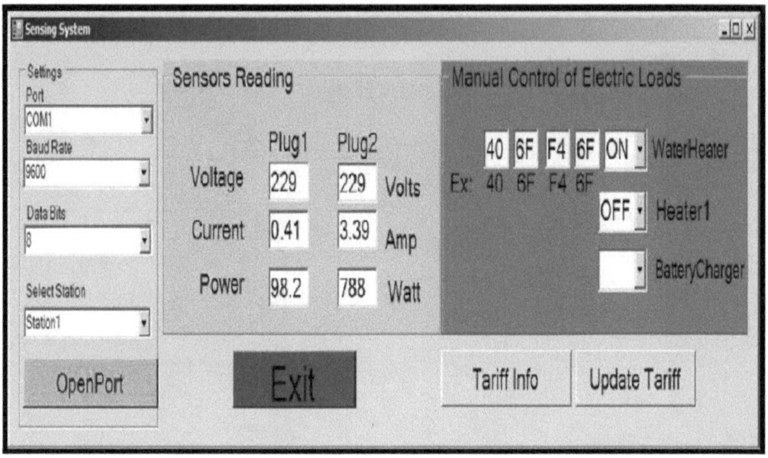

Fig. 7.42 Prototype of the smart power monitoring system and Graphical User Interface

The prototype has been tested and results achieved for certain home electrical appliances. Table 7.3 below shows the percentage error for all measured parameters with the corresponding references. From the very low percentage error of power, it can be said that power can be calculated without considering power factor.

We aim to determine the areas of daily peak hours of electricity usage levels and come with a solution by which we can lower the consumption and enhance better utilisation of electricity. At present, system is independently tested with different loads of domestic appliances. In the later stage, system will be integrated with co-systems like smart home behavior recognitions systems to determine the wellness of the inhabitant in terms of energy consumption.

Table 7.3 Percentage error of measured voltage, current and power

Appliance	Ref. Load (W)	V ref (V)	I. Ref (amps)	Mea. Vol (V)	%Error-Voltage	Mea. Cur (amps)	%Error-Current	Cal. Power (W)	%Error-Power
Bulb	25	229	0.11	229	0	0.11	0.00	25.19	0.76
Bulb	36	229	0.16	230	0.44	0.17	6.25	39.1	8.61
Bulb	59	229	0.26	229	0	0.27	3.85	61.83	4.80
Bulb	73	229	0.32	229	0	0.32	0.00	73.28	0.38
Bulb	98	228	0.43	229	0.44	0.42	2.33	96.18	1.86
Heater	401	226	1.73	225	0.44	1.82	5.20	409.5	2.12
Heater	755	223	3.41	222	0.44	3.45	5.86	781.91	3.56
Heater	1155	224	5.15	223	0.44	5.16	6.43	1145.23	0.85
Toaster	811	226	3.49	226	0	3.56	1.17	808.97	0.25
Toaster	665	234	2.90	235	0.43	2.73	2.01	658.72	0.94
Heater	733	236	3.11	237	0.42	2.91	0.19	703.1	4.08
Heater	1217	235	5.19	236	0.43	5.05	2.70	1192	2.05
Heater	1902	233	8.17	234	0.43	7.83	4.16	1869	1.74
Kettle	1995	233	8.72	233	0	8.18	6.20	1917.2	3.90

7.4 Suggested Further Reading

The readers may read the following references for more knowledge and information.

[1] Kaur, K., Mukhopadhyay, S.C., Schnepper, J., Haefke, M., Ewald, H.: A Zigbee Based Wearable Physiological Parameters Monitoring System. IEEE Sensors Journal 12(3), 423–430 (2012)

[2] Bonato, P.: Wearable sensors/systems and their impact on biomedical engineering. IEEE Eng. Med. Biol. Mag. 22(3), 18–20 (2003)

[3] Anliker, U., Ward, J.A., Lukowicz, P., Tröster, G., Dolveck, F., Baer, M., Keita, F., Schenker, E.B., Catarsi, F., Coluccini, L., Belardinelli, A., Shklarski, D., Alon, M., Hirt, E., Scmid, R., Vuskovic, M.: AMON: A Wearable Multiparameter Medical Monitoring and Alert System. IEEE Transactions on Information Technology in Biomedicine 8(4) (December 2004)

[4] `http://en.wikipedia.org/wiki/Normal_human_body_temperature`

[5] Kannus, P., Sievanen, H., Palvanen, M., Jarvinen, T., Parkkari, J.: Prevention of falls and consequent injuries in elderly people. Lancet 366(9500), 1885–1893 (2005)

[6] Goertzel, B.: Patterns of Awareness. Dynamical Psychology, http://www.goertzel.org/dynapsyc/2004/HardProblem.htm (retrieved on December 7, 2011)
[7] AlMejrad, A.S.: Human Emotions Detection using Brain Wave Signals: A Challenging. European Journal of Scientific Research 44, 640–659 (2010)
[8] Ekman, P., Sorenson, E.R., Friesen, W.V.: Pan-cultural elements in facial displays of emotions. Science 164, 86–88 (1969)
[9] Giannakopoulos, T., Pikrakis, A., Theodoridis, S.: A dimensional approach to emotion recognition of speech from movies. In: IEEE International Conference on Acoustics, Speed and Signal Processing (ICASSP), Taipei, pp. 65–68 (2009)
[10] Camurri, A., Lagerlof, I., Volpe, G.: Recognizing emotion from dance movement: comparison of spectator recognition and automated techniques. International Journal of Human-Computer Studies 59, 213–225 (2003)
[11] Healey, J.A., Picard, R.W.: Detecting stress during real-world driving tasks using physiological sensors. IEEE Transactions on Intelligent Transportation Systems 6(2), 156–166 (2005)
[12] Sierra, A.S., Ávila, C.S., Casanova, J.G., Pozo, G.B.: A stress detection system based on physiological signals and fuzzy logic. IEEE Transactions on Industrial Electronics, 4857–4865 (October 2011)
[13] Octavian, A.P., Pedro, M.B., Silva, G., Joaquim, M., Eduardo, C.P., Gabriela, P.: Physiological Parameters Measurement Based on Wheelchair Embedded Sensors and Advanced Signal Processing. IEEE Transactions on Instrumentation and Measurement 59(10), 2564–2574 (2010)
[14] Thagard, P.: How to make decisions: coherence, emotion, and practical inference. In: Milligram, E. (ed.) Varieties of Practical Inference. MIT Press, Cambridge (2001)
[15] Damasio, A.R.: Descartes' error: emotion, reason, and the human brain. G.P, Putnam's Sons, New York (1994)
[16] Prinz, J.: Which emotions are basic? In: Evans, D., Cruse, P. (eds.) Emotion, Evolution, and Rationality. Oxford University Press (2004)
[17] Fredrickson, B.L.: What good are positive emotions? Review of General Psychology 2, 173–186 (1998)
[18] Gross, J.J.: The emerging field of emotional regulation: An integrative review. Review of General Psychology 2, 271–299 (1998)
[19] Shelley, K.H.: Photoplethysmography: Beyond the Calculation of Arterial Oxygen Saturation and Heart Rate. International Anesthesia Research Society 105(6), S31–S36 (2007)
[20] Zakrzewski, M., Raittinen, H., Vanhala, J.: Comparison of Center Estimation Algorithms for Heart and Respiration Monitoring With Microwave Doppler Radar. IEEE Sensors Journal 12(3), 627–634 (2012)
[21] Bonds, Q., Gerig, J., Weller, T.M., Herzig, P.: Towards Core Body Temperature Measurement via Close Proximity Radiometric Sensing. IEEE Sensors Journal 12(3), 519–526 (2012)
[22] A Tutorial on Clustering Algorithms. K-means clustering, http://home.dei.polimi.it/matteucc/Clustering/tutorial_html/kmeans.html (retrieved on December 11, 2011)
[23] The University of Waikato, Weka, http://www.cs.waikato.ac.nz/ml/weka/ (retrieved on September 17, 2011)

[24] Li, L., Hu, X., Jian, H., Ketai, H.: Design of new architecture of AMR system in Smart Grid. In: Proceedings of the 6th IEEE Conference on Industrial Electronics and Applications (ICIEA) 2011, pp. 2025–2029 (2011)

[25] Han, D.-M., Lim, J.-H.: Smart home energy management system using IEEE 802.15.4 and zigbee. IEEE Transactions on Consumer Electronics 56(3), 1403–1410 (2010)

[26] Andrey, E., Morelli, J.: Design of a Smart Meter Techno-Economic Model for Electric Utilities in Ontario. In: Proceedings of the IEEE- Electric Power and Energy Conference (EPEC), pp. 1–7 (2010)

[27] Benzi, F., Anglani, N., Bassi, E., Frosini, L.: Electricity Smart Meters Interfacing the Households. IEEE Transactions on Industrial Electronics 58(10) (2011)

[28] Kunold, I., Kuller, M., Bauer, J., Karaoglan, N.: A system concept of an energy information system in flats using wireless technologies and smart metering devices. In: Proceedings of the IEEE 6th International Conference on Intelligent Data Acquisition and Advanced Computing Systems (IDAACS), pp. 812–816 (2011)

Kolibri 360 GmbH
Ferdinand-Jühlke-Straße 7
99095 Erfurt, Deutschland
produktsicherheit@kolibri360.de